THE AURORA
Sun–Earth Interactions
Second edition

WILEY-PRAXIS SERIES IN ASTRONOMY AND ASTROPHYSICS
Series Editor: John Mason, B.Sc., Ph.D.

Few subjects have been at the centre of such important developments or seen such a wealth of new and exciting, if sometimes controversial, data as modern astronomy, astrophysics and cosmology. This series reflects the very rapid and significant progress being made in current research, as a consequence of new instrumentation and observing techniques, applied right across the electromagnetic spectrum, computer modelling and modern theoretical methods.

The crucial links between observation and theory are emphasised, putting into perspective the latest results from the new generations of astronomical detectors, telescopes and space-borne instruments. Complex topics are logically developed and fully explained and, where mathematics is used, the physical concepts behind the equations are clearly summarised.

These books are written principally for professional astronomers, astrophysicists, cosmologists, physicists and space scientists, together with post-graduate and undergraduate students in these fields. Certain books in the series will appeal to amateur astronomers, high-flying 'A'-level students, and non-scientists with a keen interest in astronomy and astrophysics.

ROBOTIC OBSERVATORIES
Michael F. Bode, Professor of Astrophysics and Assistant Provost for Research, Liverpool John Moores University, UK

THE AURORA: Sun–Earth Interactions
Second Edition
Neil Bone, British Astronomical Society and University of Sussex, Brighton, UK

PLANETARY VOLCANISM: A Study of Volcanic Activity in the Solar System, Second edition
Peter Cattermole, formerly Lecturer in Geology, Department of Geology, Sheffield University, UK, now Principal Investigator with NASA's Planetary Geology and Geophysics Programme

DIVIDING THE CIRCLE: The Development of Critical Angular Measurement in Astronomy 1500–1850
Second edition
Allan Chapman, Wadham College, University of Oxford, UK

THE DUSTY UNIVERSE
Aneurin Evans, Department of Physics, University of Keele, UK

MARS AND THE DEVELOPMENT OF LIFE, Second edition
Anders Hansson, Ph.D., Senior Science Consultant, International Nanobiological Testbed

COMET HALLEY - Investigations, Results, Interpretations
Volume 1: Organization, Plasma, Gas
Volume 2: Dust, Nucleus, Evolution
Editor: John Mason, B.Sc., Ph.D.

ELECTRONIC IMAGING IN ASTRONOMY: Detectors and Instrumentation
Ian. S. McLean, Department of Astronomy, University of California at Los Angeles, California, USA

URANUS: The Planet, Rings and Satellites
Ellis D. Miner, Cassini Project Science Manager, NASA Jet Propulsion Laboratory, Pasadena, California, USA

THE PLANET NEPTUNE: An Historical Survey Before Voyager, Second edition
Patrick Moore, CBE, D.Sc.(Hon.)

ACTIVE GALACTIC NUCLEI
Ian Robson, Director, James Clerk Maxwell Telescope, Director Joint Astronomy Centre, Hawaii, USA

THE HIDDEN UNIVERSE
Roger J. Tayler, Astronomy Centre, University of Sussex, Brighton, UK

THE AURORA

Sun–Earth Interactions
Second edition

Neil Bone
British Astronomical Association
and University of Sussex

JOHN WILEY & SONS
Chichester • New York • Brisbane • Toronto • Singapore

Published in association with
PRAXIS PUBLISHING
Chichester

First published in 1991
This Second edition published in 1996 by
John Wiley & Sons Ltd
in association with Praxis Publishing Ltd

Wiley Editorial Offices

John Wiley & Sons Ltd, Baffins Lane,
Chichester, West Sussex PO19 1UD, England

John Wiley & Sons, Inc., 605 Third Avenue,
New York, NY 10158-0012, USA

Jacaranda Wiley Ltd, G.P.O. Box 859, Brisbane
Queensland 4001, Australia

John Wiley & Sons (Canada) Ltd, 22 Worcester Road,
Rexdale, Ontario M9W 1L1, Canada

John Wiley & Sons (Asia) Pte Ltd, 2 Clementi Loop #02-01,
Jin Xing Distripark, Singapore 0512

A catalogue record for this book is available from the British Library

ISBN 0-471-96023-3 Cloth
ISBN 0-471-96024-1 Paperback

Printed and bound in Great Britain by Hartnolls Ltd, Bodmin

For Gina, with love

Table of contents

The colour plate section appears between pages 76 and 77

List of illustrations

COLOUR PLATES

LIST OF TABLES

Preface

The popular literature has, over the years, been rather unkind to phenomena of astronomical origins occurring in the Earth's atmosphere. Together with meteors, the aurora has usually been presented as something rather exotic and not really astronomical anyway, before the author passes on to the next topic. It is my hope, here, to redress the balance somewhat, by presenting an account of the aurora, its causes and related phenomena, in a form accessible to the reasonably well-informed amateur astronomer.

Many years ago, as a young newcomer to astronomy, I was an avid reader of the numerous general introductions to the subject. Among these, one of the most eye-catching features would often be a garishly coloured 'artist's impression' of the aurora borealis, or northern lights. These displays, I was frequently told, could only be seen from the high Arctic latitudes, and were caused by a rain of particles into the atmosphere from the Van Allen belts girdling the Earth. Little seems to have changed in the past 25 years, and modern beginners can anticipate being similarly misinformed!

Presumably, the majority of the artists responsible for the illustrations had never seen a display of the aurora: if they had, I might not have mistaken my first display, in 1973, for an unexpectedly early moonrise! One problem was that the lights I was witnessing in the sky bore absolutely no resemblance to the perfectly symmetrical and strangely hued paintings in the popular astronomy books. Another was that these illustrations gave no impression of scale. The aurora can assume many forms, of which I have striven to provide photographic examples in the atlas section here. Familiar star groups, such as the Plough or Cassiopeia, should help give some idea of the extent which these displays can have.

I was fortunate enough to have been born and raised in Scotland, from where I enjoyed numerous subsequent opportunities to observe the aurora (not a strictly polar phenomenon after all!). It seemed natural, therefore, to try and learn more about the aurora and its causes. Beyond the occasional cursory mention in general texts, and *slightly* less cursory mention in books about the Sun, however, there was little to be found in the easily accessible literature. It seemed curious that texts are readily available on other popular (and not-so-popular) observational topics such as variable stars, deep sky objects and planets, but not about events going on more or less straight above us! This book is an attempt to fill the gap.

Certainly the aurora merits more than just a passing glance. It is one of the most awesome of all natural sights, once seen, never forgotten, as the many who saw the huge—more or less global—display of March 1989 will testify. There remain some valuable contributions to be made on the observational front by amateur astronomers, in the fields both of aurora and of noctilucent cloud work: I have attempted to outline these in the later chapters. It goes without saying that amateur observations of that other atmospheric astronomical phenomenon, meteors, remain of great value, too.

Having moved, through professional commitments, to the south of England, I now find the aurora, sadly, to be an infrequent visitor to my skies. There was a certain degree of personal irony when, on the night of March 24 last, one of the best displays for a couple of years—rays extending to Polaris and red patches of auroral light filling the eastern sky—was seen even from here in Sussex, in the week when I completed the text of this book!

To a large degree, this event demonstrated the unpredictable nature of the aurora: few anticipated such an early revival of auroral activity from the lull which usually appears to accompany sunspot maximum. We can hope for several further such events penetrating to lower latitudes in the times ahead. It is my hope that what follows will allow those who see these events to better understand and observe them.

Chichester
March 31 1991

Neil Bone

Preface to the second edition

The publication of the first edition of this book proved more timely than might originally have been anticipated, coinciding with the vigorous season for aurorae in the autumn of 1991 as activity reached its secondary peak. The stunning, bright and active aurora of 1991 November 8–9, particularly, did much to bring the aurora to the attention of amateur astronomers and others worldwide. The slightly unusual causes behind this event were an almost immediate prompt that more needed to be brought into the existing text regarding the role of disappearing filaments in generating at least some major aurorae. In the interim, much has also been done by professional workers to perhaps revise our views of how aurorae come about—the very Sun–Earth interactions of my sub-title. In particular, the role of coronal mass ejections (CMEs) in causing aurorae, and the inter-relation between CMEs and solar flares, has come very much to the fore in professional debate. I have tried to give some flavour of these changing ideas in the chapter on solar activity here.

In the time since the first edition, much more has also been learned about the interactions of other bodies with the solar wind, and about the nature of the outer solar atmosphere within which Earth orbits. The need to bring the appropriate sections up to date has now hopefully been met.

I have taken on board some reviewers' suggestions following the first edition, most particularly in adding a glossary, which many felt would be useful.

The section on noctilucent clouds has been retained and slightly extended. Amateur astronomers in the northern United States and Canada are becoming increasingly aware of the possibility and value of observing these enigmatic high-atmosphere clouds, recording of which is an integral part of the work of the BAA Aurora Section.

As we head towards 1996 and the next sunspot minimum, auroral activity has dwindled. The onset of the next sunspot cycle has already been detected at professional observatories, and it is likely that by 1997–98, we shall see a marked upturn in activity, including—as in the past two cycles—auroral displays visible to lower latitudes. It is my hope that this revised and extended volume will provide a useful guide to the aurora and related phenomena for those whose interest will inevitably be stirred by events in the rise towards the next solar maximum, at the end of the century.

Chichester
November 5 1995

Neil Bone

Acknowledgements

Many people have helped me greatly in compiling this account of the aurora. I should like to thank, particularly, Ron Livesey and Dave Gavine of the British Astronomical Association Aurora Section for reading, and commenting upon, sections of the text, and for suggesting several possible improvements. Dave also generously offered a free run of his unparalleled slide collection, from which several fine examples of auroral forms were taken. Richard Pearce and Tom McEwan also kindly allowed use of their excellent photographic material.

Tony Kinder, Director of the BAA Historical Section helped me to fill in more detail on the auroral writings of Gregory of Tours. Many reviewers of the first edition provided comments which have benefited the preparation of the current volume. In particular, I am grateful to Dr Michael Gadsden in this respect, for gently making me aware of the shortcomings where necessary.

During the preparation of the first edition, of the book, John Mason (who talked me into it, in the first place) was a great help, offering useful suggestions on how to proceed during numerous productive evenings in the estimable Murrell Arms in Barnham. In coming to prepare this second edition, I have again been guided by John's suggestions.

I am especially grateful to Dr William Kurth of the University of Iowa, who carefully read through the first draft of the manuscript for this edition, bringing to light a number of points where revisions, re-phrasing or additions were necessary, and providing many positive suggestions. Any errors and omissions which remain are solely the responsibility of the Author.

Two-and-a-bit-year-old Miranda deserves mention for carefully hiding the disks on which I was originally working last August, forcing me to break out of a season of sloth during the particularly hot summer just past. I have every confidence that these will turn up, now that the work has been re-done!

Last, and by no means least, my wife Gina deserves a special 'thankyou' for acting as guinea pig for many of the less readable prototypes of some sections, and for patiently putting up with the scatter of papers around the house during the writing of this volume! As promised in the first edition, the kitchen was duly painted; I await instructions as to my next DIY exercise!

1

Introduction

1.1 ASTRONOMICAL PHENOMENA IN THE EARTH'S HIGH ATMOSPHERE

To many astronomers, the atmosphere of the Earth is nothing more than a nuisance, its turbulence and grime hindering precise observations. Amateur astronomers in maritime temperate locations such as the British Isles are all too familiar with the difficulty of attempting extended observing programmes when frontal weather systems and their attendant cloud and rain are never far away. Perhaps the only compensation for the loss of critical nights when rare astronomical events might occur is given by the spectacular optical effects sometimes to be seen in the advancing cloudsheets ahead of the front by the assiduous observer (Greenler, 1989; Minnaert, 1954).

Professional observatories are often sited on high mountains in remote locations in order to get above much of the haze and unsteady seeing of the lower atmosphere. Even such measures may be insufficient, however, for the astronomer wishing to observe in wavelengths of the electromagnetic spectrum which are unable to penetrate the atmosphere. To observe in ultraviolet and other wavelengths absorbed by the atmosphere, it is necessary to send telescopic equipment into space, aboard satellites such as the International Ultraviolet Explorer or Hubble Space Telescope, with concomitant additional expense and further inaccessibility.

Sensitive equipment in near-Earth orbit may still be at the mercy of atmospheric interference: observations in certain ultraviolet wavelengths from the Space Shuttle have been severely hindered by the phenomenon of 'Shuttle Glow' (Hunton, 1990), resulting from the impact on spacecraft surfaces of the orbital wind generated by the Shuttle's passage through the outer fringes of the Earth's atmosphere. Probably, the only complete solution to all the problems presented for precise astronomical observations by the Earth's atmosphere will be the eventual establishment of observatories on the Moon (Burns *et al.* 1990).

Life on Earth is, of course, utterly dependent on the atmosphere. The very absorption which thwarts ground-based observations at ultraviolet wavelengths is crucially important in reducing the damaging effects of this component of sunlight to organisms on the Earth's surface. In the absence of the stratospheric ozone layer, which initially appeared as a con-

sequence of the evolution of oxygen-liberating photosynthesis by green plants two billion†
years ago, all life on our planet would still be restricted to the aquatic environment.

As well as providing the necessary medium for supporting life on the planet, Earth's
atmosphere is a location in which several phenomena of astronomical interest in their own
right do take place. The atmosphere provides a shield against the continual bombardment
of the Earth by small particles of interplanetary debris (meteoroids), most of which are
completely ablated upon impact at high velocities of anywhere between 11 and 72 km s^{-1}.

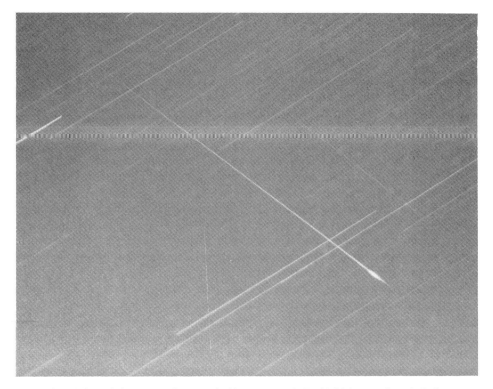

Fig. 1.1. Several phenomena of astronomical interest occur in Earth's high atmosphere, including
the aurora. A very productive area for both amateur and professional observers is the study of
meteor activity. In addition to assessing visual rates, amateur astronomers can make a useful
contribution by attempting photography of meteors, to obtain positional information. Time
exposures with fast black and white film give the best results. In this time exposure with a
stationary camera, the stars trailed gradually across the field as a result of the Earth's rotation.
Cutting across the star trails is the much longer trail left by a bright Perseid meteor, recorded on
1994 August 13–14 at 0110 UT by the Author.

Observations by astronomers, both professional and amateur, of the *meteor* phenomena
which result from these atmospheric impacts have contributed to knowledge of the nature
of the interplanetary dust (Bone, 1993). Much of the meteoric material which enters the
atmosphere is of cometary origin, giving rise to the regular annual meteor showers such as
the Perseids or Orionids. Amateur astronomers continue to carry out useful work by con-
ducting simple naked-eye visual meteor watches. Systematic observations of the annual

† Throughout this book, a billion is a thousand million.

showers reveal much about how meteor streams evolve through their interactions with solar radiation and gravitational perturbations by the planets over long periods.

Occasional fragments of asteroidal material, deflected into Earth-crossing orbits by collisions in the main belt between Mars and Jupiter, can survive atmospheric passage to be picked up on the ground for direct study. It is possible through such work to examine some of the most primitive material left over from the formation of the solar system, and make deductions about the origin of the Earth and other planets. Early recovery of certain meteoritic types is particularly desirable. Patrol photography to detect bright, potentially meteorite-dropping meteors (fireballs) is a valuable and somewhat neglected field of work which can be, and has been, carried out by amateur astronomers operating simple camera systems.

Some of the debris resulting from meteoric phenomena remains suspended in the high atmosphere for considerable periods, where it may provide condensation nuclei for another phenomenon of interest, *noctilucent clouds*. These 'night-shining' clouds, seen during the summer months in either hemisphere of the Earth, form at great heights (around 85 km), above 99% of the atmosphere. They are believed to comprise water-ice condensed around small nuclei, which are most likely of meteoric origin, but may also come from volcanic or even ionospheric sources. Noctilucent clouds lie on the fringe of space, and also fall into a fringe area between atmospheric and astronomical study. Observations of noctilucent clouds have been collected over many years since the 1880s, and may sensitively indicate changes—possibly man-made—in the nature of the atmosphere as a whole. Their study is included in the field of aeronomy, which encompasses phenomena of the middle and high atmosphere.

By far the most spectacular of the astronomically related phenomena of the high atmosphere are the displays of the aurora. The *aurora borealis* or *aurora australis* are, respectively, more or less permanent features of the northern and southern polar night, being distributed in oval regions around either geomagnetic pole (section 6.3.1). The aurora borealis is often seen by travellers on high-latitude transatlantic flights between America and Europe. On occasion, particularly during periods of intense activity on the Sun, the aurorae can extend to lower latitudes, becoming visible from the more populous parts of the world. Sadly, most general astronomical texts treat the aurora purely as a phenomenon of very high latitudes, and pass over it in a very few paragraphs, if they even mention it at all. This disregards the fact that aurorae can be seen with surprising frequency by the attentive observer at the latitudes of, say, the northern United States, northwest Europe, or New Zealand.

Observations of the aurora reveal much about conditions in the Earth's high atmosphere, and in the interplanetary medium which surrounds the Earth. The solar system is pervaded by the continually expanding tenuous outer atmosphere of the Sun, which carries with it magnetic fields whose strength and direction are determined by features of the solar surface such as sunspots, or by coronal holes rooted in the inner solar atmosphere.

Confirmation of the existence of the predicted *solar wind* (as the expansion of the solar atmosphere is now known) was one of the early triumphs of the Space Age. A more precise understanding of processes in the solar wind will be an essential next step in further exploration of the solar system, particularly if manned endeavours such as projected Mars missions are to be considered. While the atmosphere provides a shield against incoming meteoric material and cosmic rays, the magnetosphere—the region of near-Earth space dominated by Earth's magnetic field—plays a less immediately obvious role in deflecting most of the energetic particles of the solar wind around the Earth. Spacecraft in the inter-planetary environment enjoy no such shielding, and the effects of long-term exposure to high-energy particles and radiation on living organisms will have to be better understood before manned interplanetary missions can be contemplated.

Even within the shield provided by the terrestrial magnetic field, regions exist where subatomic particles—electrons and heavier ions—can become trapped, spiralling back and forth along field lines. This motion can result in the particles becoming accelerated to high energies. Spacecraft traversing these trapping regions (the Van Allen belts) are, con-sequently, subjected to increased radiation dosage, which may cause damage to delicate components such as solar panel arrays. 'Bit flips' resulting from the passage of energetic particles through computer memory chips can also give rise to problems.

In particular, the region of the South Atlantic Anomaly, where the inner Van Allen belt dips close to the Earth, 250 km above the ocean off the coast of Brazil, is a hazard for satellites in low-Earth orbit. It is desirable for orbiting satellites to spend as little time as possible within the Van Allen belts (Sherrill, 1991).

Immersion of the Earth within pockets of southerly directed magnetic field in the solar wind can give rise to the often spectacular displays of the northern and southern lights (seen simultaneously in either hemisphere), as particles enter the magnetosphere and are accelerated into the high atmosphere. The accompanying weakening of the global mag-netic field carries the aurorae out of their normal, undisturbed high latitude zones of occur-rence, and towards the equator. Those living at moderately high temperate latitude loca-tions, such as those of northwest Europe or the northern United States can therefore see the aurora quite frequently, especially at times of high solar activity.

1.2 INFLUENCE OF AURORAL PHENOMENA ON HUMAN ACTIVITIES

It is important to realize that, for all their poetic visual beauty, the phenomena associated with the arrival of energetic particles from the Sun are also of increasing relevance to human affairs. During one particularly active auroral storm at the end of 1989 (a remark-able year for the occurrence of aurorae at lower latitudes), for example, the cosmonauts aboard the Soviet *Mir* space station reputedly accumulated the maximum recommended dose of radiation exposure for an entire year. Throughout the same evening, ground-based observers in Hungary at latitude 47°N witnessed extensive red 'clouds' of aurora across their skies. In certain areas of the western press, which really should have known better, reports of these sightings were exemplified as revealing new-found honesty regarding the existence of UFOs under the transforming politics of *glasnost* in Eastern Bloc countries!

Aurorae must, indeed, account for a substantial fraction of those few UFO reports in the western world which cannot otherwise be ascribed to sightings of the planet Venus, bright stars twinkling vigorously in the unsteady atmosphere near the horizon, or bright meteors. The cigar-shaped green pulsing 'spacecraft', frequently mentioned by von Daniken and other UFO advocates, more probably correspond to auroral arc forms, while other cruci-form objects are likely to be manifestations of coronal aurora.

Transport and communication are also influenced by auroral activity. Many ship navigation systems use signals transmitted from satellites to accurately determine position at sea. During a week of intense auroral activity in February 1982, the resulting strong electrical currents flowing at orbital heights disabled the Marecs-B marine navigation satellite. Gyro-stabilizers aboard the Canadian communications satellites Anik E1 and Anik E2 were both affected by a geomagnetic disturbance in January 1994.

Surges in electrical grid systems at ground level have resulted in power failures during auroral activity for residents at high latitudes, such as those of northern Canada and Scandinavia. Notably, the Great Storm of 1989 March 13–14 caused extensive electrical blackouts in the Quebec Province of Canada.

Auroral ground currents were also perceived as a threat to the Alaska oil pipeline, which has had to be specially protected against any resulting corrosion. Corrosion due to such currents has been cited as the cause of a disastrous gas pipeline explosion close to the Trans-Siberian Railway, which resulted in more than 650 deaths in June 1989 (Penman, 1995).

The violent solar events which trigger more vigorous auroral activity, and activity penetrating to lower latitudes, may be accompanied by other effects on the terrestrial environment. Strong ultraviolet emission during solar flares, for example, causes an increase in the ionization of the atmosphere 80 km above the Earth's surface. A result of this can be disruption of short-wave radio communication.

Fig. 1.2. An auroral display imaged from low Earth orbit from the Space Shuttle. The aurora is seen well clear of the curved limb of the Earth at the left of the image, in which the sharp lower boundary of the aurora at around 100 km altitude is particularly marked. NASA photograph.

At times of high solar activity, Earth's outer atmosphere becomes more extended as a result of heating produced by an enhanced flux of X-rays and ultraviolet radiation from the Sun due to solar flares. One consequence of this is an increase in the 'drag' experienced by artificial satellites, particularly those in lower orbits. Such objects may therefore decay from orbit earlier than might be desirable.

Perhaps the best-known victim of this effect was NASA's Skylab, which re-entered prematurely in July 1979. It had originally been planned to use the Space Shuttle to boost Skylab back to a higher 'safe' orbit, but this was thwarted by delays in American space programme development and high levels of sunspot activity in the run-up to the 1980 solar maximum.

Another celebrated loss resulting from increased atmospheric density at orbital heights during high sunspot activity was, ironically, the Solar Maximum Mission (SMM) satellite. Popularly known as 'Solar Max', the satellite was launched in 1980, and was the first to be retrieved and repaired in orbit, by astronauts aboard the Space Shuttle *Challenger* in 1984. SMM decayed from orbit in April 1990. Observations obtained using both Skylab and the Solar Maximum Mission added greatly to understanding of processes in the near-solar environment.

Spacecraft involved in planetary exploration are, from time to time, exposed to increased fluxes of energetic subatomic particles ejected into interplanetary space during the same violent solar events which can trigger terrestrial aurorae. It is essential, if probes such as Galileo or Cassini are to function successfully, that delicate computer and other equipment be 'hardened' against possible damage resulting from collisions with such particles. The provision of adequate back-up hardware and software is vitally important to the planning of unmanned spacecraft missions which are likely to undergo prolonged exposure to the variable conditions of the interplanetary medium (Williams, 1990).

Energetic particles arriving in near-Earth space following violent solar events are, potentially, hazardous to astronauts working in the orbital environment. As such operations become increasingly routine, a better understanding of the underlying mechanisms which produce the aurora and related phenomena will also become important: forecasts of the 'weather' in near-Earth space might be as necessary to orbital industrial activities based around projected space stations as are gale warnings to deep-sea fishermen!

1.3 AURORAL FOLKLORE

The aurora has, not surprisingly, often entered into the folklore of peoples living at higher latitudes. The Scottish city of Aberdeen is connected to the northern lights or 'Heavenly Dancers' in a popular song, for example. Robert Burns, Scotland's national poet, makes mention of the aurora in his galloping epic *Tam O'Shanter*:

> But pleasures are like poppies spread,
> You seize the flow'r, its bloom is shed,
> Or like the snow falls in the river,
> A moment white —then melts for ever,
> Or like the borealis race,
> That flit ere you can point their place;

Or like the rainbow's lovely form
Evanishing amid the storm. —
Nae man can tether time nor tide;
The hour approaches Tam maun ride

The Edinburgh Edition, 1793

The much-travelled poet, Robert William Service (1874–1958) spent some time in the far north of Canada, during the declining years of the late-nineteenth-century Klondike and Yukon gold rush. Some of his poetry (for example, Service 1911a, 1911b) provides a fascinating record of frontier life. Not surprisingly, the aurora occasionally features as a backdrop:

There where the mighty mountains bare their fangs unto the moon;
There where the sullen sun-dogs glare in the snow-bright, bitter noon,
And the glacier-gutted streams sweep down at the clarion call of June:

There where the livid tundras keep their tryst with the tranquil snows;
There where the silences are spawned, and the light of hell-fire flows
Into the bowl of the midnight sky, violet, amber and rose

There where the rapids churn and roll, and the ice flows following run;
Where the tortured, twisted rivers of blood rush to the setting sun—
I've packed my kit and I'm going, boys, ere another day is run.

The Heart of the Sourdough.

A longer narrative tells the tale of three miners seeking their fortune in the far north, guided by their dreams. The sole survivor—a down-and-out—relates their experiences and describes the aurora:

Oh, it was wild and weird and wan, and ever in camp o' nights
We would watch and watch the silver dance of the mystic Northern Lights.
And soft they danced from the Polar sky and swept in primrose haze;
And swift they pranced with their silver feet, and pierced with a blinding blaze.
They danced a cotillion in the sky; they were rose and silver shod;
It was not good for the eyes of man—'twas a sight for the eyes of God.
It made us mad and strange and sad, and the gold whereof we dreamed
Was all forgot, and our only thought was of the lights that gleamed.

And the skies of night were alive with light, with a throbbing thrilling flame;
Amber and rose and violet, opal and gold it came.
It swept the sky like a giant scythe, it quivered back to a wedge;
Argently bright, it cleft the night with a wavy golden edge.
Pennants of silver waved and streamed, lazy banners unfurled;
Sudden splendors of sabres gleamed, lightning javelins were hurled.
There in our awe we crouched and saw with our wild, uplifted eyes
Charge and retire the hosts of fire in the battle-field of the skies.

The storyteller informs us that the aurora originates from a hollow mountain range on the polar rim. Echoing a popular belief among the frontier gold miners of the time, he finally reveals:

> Some say that the Northern Lights are the glare of the Arctic ice and snow;
> And some say that it's electricity, and nobody seems to know.
> But I'll tell you now—and if I lie, may my lips be stricken dumb—
> It's a *mine*, a mine of the precious stuff that men call radium.
>   *The Ballad of the Northern Lights.*

In much earlier times, Norse mythology makes frequent reference to Bifrost's bridge, a burning, trembling arch across the sky, over which the gods could travel from Heaven to Earth. It is not unlikely that the inspiration for the bridge was the aurora. Similarly, the vivid red sometimes seen in intense auroral displays can probably be associated with the Viking 'vigrod', or war-reddening (Egeland and Brekke, 1984). In some traditions, auroral rays were perceived as lights carried by the Valkyries as they rode the sky. In a parallel to Bifrost's bridge, Finnish mythology refers to a river—Rutja—which stood in fire, and marked the boundary between the realms of the living and the dead.

In Norwegian folklore, the aurora has been described as a harbinger of harsh weather: snow and wind are believed to follow bright displays. Another Norwegian folk-legend suggests that the aurora is a celestial dance by the souls of dead maidens.

Eskimo peoples in the Hudson Bay area of North America, and elsewhere, are naturally very much aware of auroral phenomena. A common belief among the Eskimos is that the aurora can be attracted by whistling to it, while a handclap will cause it to recede. Other Eskimo beliefs suggest that the aurora is produced by spirits, playing a game of celestial football with the skull of a walrus. (One group, on Nanivak Island, suggested that a human skull was, instead, used by walrus spirits!)

Some Eskimo groups regard the aurora as an indicator of good weather to be brought by the spirits. Alaskan Eskimos at Point Barrow saw the aurora as malevolent, and carried weapons for protection if venturing outside when it was present (Ray, 1979). It is also said by some Eskimos that: 'He who looks long upon the aurora soon goes mad!'

Some tribes of North American Indians believed the aurora to be the light of lanterns carried by spirits seeking the souls of dead hunters. Like the Point Barrow Eskimos, Fox Indians in Wisconsin feared the aurora, seeing in it the ghosts of their dead enemies. Other tribes perceived the aurora as the light of fires used by powerful northern medicine men.

The aurora has also entered the folklore of the Australian aborigines, who saw it as the dance of gods across the sky. To the Maoris of New Zealand, the aurora is *Tahu-Nui-A-Rangi*, the great burning of the sky.

Aurorae may well have been the source of Chinese dragon legends. The twisting snake-like forms of active auroral bands are often portrayed as celestial 'serpents' in ancient chronicles. European dragon legends, too, may have their origin in auroral activity. How ironic that the English patron saint (St George) is now thought by many researchers to have had his legendary battle with that most Scottish of phenomena, the aurora, rather than a dragon!

In ancient Roman and Greek records, references may sometimes be found to 'chasmata' in the sky, the auroral arc structure being regarded in such instances as being the mouth of

Fig. 1.3. Folded multiple rayed bands of auroral activity are a frequent sight in the night skies of high latitudes during the breakup phase of substorm activity (section 6.4.3). The heavenly dance of the aurora has been the source of many Norse and Native American legends.

a celestial cave. The term *isochasms* is used nowadays to relate two geographical points which share an identical frequency of auroral occurrence.

In more modern times, misconceptions regarding the cause of the aurora are still common among the general public. The author of a classic popular article on the aurora (Gartlein, 1947) relates how, in his youth, people in mid-west America widely believed the aurora to be the reflection of sunlight from the polar ice, disregarding the perpetual darkness of the winter months at Arctic latitudes! Another romantic notion, long-since safely dismissed, was that auroral light results from icebergs crashing together in the polar seas.

Modern scientific understanding of the processes underlying the aurora is now sufficiently advanced that good working models to describe the causes of the polar lights are available. Many fine details remain to be investigated, and further exploration of the near-Earth environment by satellites should eventually resolve the ambiguities.

1.4 WHO NAMED THE AURORA?

The phenomenon of the aurora has been known since ancient times, and, as we have already seen in section 1.3, has been described by several names. The modern nomenclature of *aurora borealis*—northern dawn—is widely credited to the French astronomer Pierre Gassendi (1592–1655), who first used the term following an excellent display seen on 1621 September 12. Others credit the first use of the name to Galileo (1564–1642) who witnessed the same display from Venice, or jointly to Galileo and Gassendi. Perhaps of interest is the suggestion (Davidson, 1985) that the term aurora borealis may have been applied by Gregory of Tours, who predates Gassendi and Galileo by more than a millennium!

Gregory of Tours (538–594) was involved in clerical and political life, and is best known for his chronicle, *The History of the Franks*. In this can be found descriptions of various phenomena as signs from the heavens including, for example, 'the meteor which country folk call suns' (parhelia or sundogs: it was common at this time to refer to a host of atmospheric phenomena—in addition to the modern accepted definition of 'shooting stars' as meteors). A number of entries describe the aurora, too.

> While we were still hanging about in Paris portents appeared in the sky. Twenty rays of light appeared in the north, starting in the east, and then moving round to the west. One of them was longer than the others and shone high above them: it reached right up into the sky and then disappeared, and the others faded away, too. (AD 578)

> While I was staying in Carnignan, I twice during the night saw portents in the sky. These were rays of light towards the north, shining so brightly that I had never seen anything like them before: the clouds were blood-red on both sides, to the east and to the west. On a third night these rays appeared again, at about seven or eight o'clock. As I gazed in wonder at them, others like them began to shine from all four quarters of the earth, so that as I watched they filled the entire sky. A cloud gleamed bright in the middle of the heavens, and these rays were all focused on it, as if it were a pavilion the coloured stripes of which were broad at the bottom but became narrower as they rose, meeting in a hood at the top. In between the rays of light there were other clouds flashing vividly as if they were being struck by lightning. This extraordinary phenomenon filled me with foreboding, for it was clear that some disaster was about to be sent from heaven. (AD 586)

One example may have been the origin of the nomenclature:

> At this time there appeared at midnight in the northern sky a multitude of rays which shone with extreme brilliance. They came together and then separated again, vanishing in all directions. The sky towards the north was so bright that you might have thought that day was about to dawn.

As we shall see later, such records have been interpreted by some workers as a means of tracing past solar/auroral activity. The vivid descriptions often indicate just what a lasting impression some of these awesome auroral events made upon the witnesses.

REFERENCES

Bone, N. (1993) *Observer's handbook: meteors*. George Philip.

Burns, J. O., Duric, N., Taylor, G. J., and Johnson, S. W. (1990) Observatories on the moon. *Scient. Am.* **262** (3) 18–25.

Davidson, N. (1985) *Astronomy and the imagination*. Routledge & Kegan Paul.

Egeland, A., and Brekke, A. (1984) The northern light: from mystery to modern space science. *Endeavour* New Series **8** 188–193.

Gartlein, C. W. (1947) Unlocking secrets of the northern lights. *National Geographic* **XCII** 673–704.

Greenler, R. (1989) *Rainbows, halos, and glories*. Cambridge University Press.

Hunton, D. E. (1990) Shuttle Glow. In 'Exploring space', *Scientific American* Special Issue.

Minnaert, M. (1954) *The nature of light and colour in the open air*. Dover.

Penman, D. (1995) The forecasters from space. *The Independent*, November 16.

Ray, D. J. (1979) Legends of the Northern Lights. *Alaska Geographic* **6** (2) 16–19.

Service, R. W. (1911a) *Songs of a Sourdough*. William Briggs.

Service, R. W. (1911b) *Ballads of a Cheechako*. William Briggs.

Sherrill, T. J. (1991) Orbital science's 'Bermuda Triangle'. *Sky and Telescope* **81** 134–139.

Williams, J. (1990) Built to last. *Astronomy* **18** (12) 36–41.

2

The aurora in history

2.1 AURORAE IN CLASSICAL TIMES

It is extremely unfortunate that the spread of artificial light pollution, accompanying population growth, has decreased Man's general awareness of astronomical phenomena, including the aurora. Few city dwellers have ever seen the Milky Way, for example, and in many locations in the civilized world, only the very brightest stars and planets may be seen. In built-up areas, all but the very brightest aurorae will be swamped by the all-pervasive background sky glow. This was not always the case, however, and historical records show that the aurora was a phenomenon well-known in antiquity when skies were darker.

It seems likely that the ancient Greeks were aware of the occasional occurrence of aurorae, which may appear in Aristotle's *Meteorologica* as glowing clouds. Aristotle classed these together with meteors and comets as atmospheric phenomena whose origin lay in friction between hot exudates from the Earth and the innermost of the heavenly spheres (Dicks, 1970).

Another example of auroral activity from classical times is a frequently related tale of the Roman Emperor Tiberius, who in AD 37 despatched a garrison to the aid of the port of Ostia, which was thought to be in flames when a red glow was seen in the sky to the north of Rome. It appears that the soldiers spent a long night marching towards an active auroral display!

Biblical references to what appear to have been ancient aurorae may also be found. In the Book of Ezekiel, for example, a heavenly apparition is described:

> And I looked, and, behold, a whirlwind came out of the north, a great cloud, and
> a fire infolding itself, and a brightness was about it, and out of the midst thereof
> as the colour of amber, out of the midst of the fire.

Over the years there has been much speculation as to the possible astronomical nature of the Star of Bethlehem. Planetary conjunctions, novae and comets have all been proposed as the celestial events interpreted by the Biblical Magi as a sign that a new King of Israel had been born. It is quite possible that the aurora might provide a further, reasonable, alternative. Computations extrapolating the ephemerides back to the time of Christ indicate that there were no bright planetary conjunctions at the appropriate time, while searches through contemporary oriental records give no indication of a candidate comet or nova. The rare penetration of auroral activity to the latitudes of the Middle East—a once-in-a-lifetime event, perhaps—would certainly be sufficiently unusual to be noted by such watchers of the sky as the Magi. In its coronal form, the aurora may very well assume a starlike appearance, with rays and other forms radiating out from a central point.

2.2 AURORAL RECORDS FROM THE DARK AND MIDDLE AGES

Searches through historical records from the Far East (often kept for astrological, rather than astronomical, purposes) by a number of workers have provided clues to the past activity of several of the annual meteor showers (Kronk, 1988). Accounts of nights when, for example, 'more than 100 meteors flew thither in the morning' or 'countless large and small meteors flew from evening till morning' provide an insight that the Perseids, currently one of the most consistent annual showers, were active as long ago as AD 36. The evolution of some meteor showers can be traced through the broad changes in their behaviour which have occurred over the centuries. The Taurids, for example, appear to have been richer than even the modern Perseids during the Middle Ages, although they are now a weak shower, producing only very low rates. This depletion over time is consistent with existing theories on the loss of material from meteor streams due to planetary perturbations and interactions with solar radiation.

The aurora, too, may be also be found reported in historical annals, and such records have been used by Eddy and others as tracers of past sunspot activity.

The aurora is described in the Norse chronicle, *The King's Mirror*, written around AD 1250:

> [The northern lights] resemble a vast flame of fire viewed from a great distance. It also looks as if sharp points were shot from this flame up into the sky, these are of uneven height and inconstant motion, now one, now another darting highest; and the light appears to blaze like a living flame.

Typical records of auroral activity from more southerly parts of Europe in the Dark and Middle Ages refer to battles fought across the sky with fiery swords or lances, and ships sailing in the heavens. The intense crimson which may often accompany active mid-latitude aurorae was often equated by contemporary witnesses with blood shed over the firmament. David Gavine, a student of the astronomical history of Scotland, has unearthed some remarkable examples:

> AD 93: Mony birnard spieris apperit, shottand in Ye air. Ane grete part of Callandair Wode semyt birnand all nicht, and na thing appering Yerof in Ye day. Ane grete noumer of schippis wer seen in Ye air.
>
> Hector Boece

> AD 352: In the nicht apperit mony swerdis and wappinis birnand in Ye aire. At last thai ran all togidder in ane grete bleiss, and evanyst out of sicht.
>
> Hector Boece

AD 839: Offt times wes sene in Ye nicht ane fyry ordinance of armit men rusching togidder with speres in Ye air, and quhen Ye tane of thame was winscust, Ye tothir sone evanist.

Hector Boece

The chronicler Boece (pronounced 'boyce') was the first Principal of King's College, Aberdeen University. Born in Dundee in 1465, he produced in 1526 his *Scotorum Historia ab Ilius Gentis Origine*, which was later translated by Bellenden. Some of the reported incidents appear to be records of auroral activity. The time around the turn of the eleventh to twelfth centuries AD appears to have been particularly rich for aurorae in mid-latitudes. Chinese records of naked-eye sunspots appear to corroborate this, such spot-groups being typically the most productive of extensive auroral displays.

Chinese and Korean records of apparent auroral activity from this period may also be found (Zhang, 1985). A typical example is:

AD 1141; ... at night, a red vapour appeared on the night sky, then two other strips of white vapour penetrating through the north pole and vicinity appeared also, sometimes they disappeared and then reappeared again.

Korean records often describe the aurora as a 'fire-like vapour'.

Further examples of auroral records may be found elsewhere in European medieval chronicles. *The Anglo-Saxon Chronicle* contains references to aurorae, meteors, comets and other celestial events (Kinder, 1990). Among these accounts are to be found references to the sky burning, and dire forewarnings over Northumbria—fiery dragons in the air. Further English records of aurorae may also be found:

AD 1235: In North England, appeared coming forth of the earth companies of armed men on horseback, with spear, shield, sword and benners displayed, in sundry forms and shapes, riding in order of battle and encountering together there. The people of the country beheld them afar off, with great wonder. (Seen for several days).

Holinshed's Chronicle

AD 1254: Seen by the monks of St Albans: in a clear night ... there appeared in the element the perfect form and likeness of a mighty great ship ... at length it seemed as the boards and joists thereof had gone in sunder, and so it vanished away.

Holinshed's Chronicle

Anyone who witnessed the awesome auroral storm covering much of the globe in March 1989 from a reasonably dark location away from artificial light pollution will have little trouble in understanding how the then unexplained phenomenon could inspire terror in the Dark Ages mind. It must have seemed, on occasion, that the world's end was imminent as the heavens blazed!

Aurorae continued to be seen through the period right up to the invention of the telescope. Further striking examples may be found of accounts in the literature well into the sixteenth century:

AD 1529: In the moneth August was seine vpon the mountaines of Striuiling afore the sone ryseng lyk fyrie candles streimes of fyre spouting furth, in the air als war sene men in harness courageouslie inuading ilk other, and sik wondiris, quhilkes with terrable feir opprest the myndes of mony.

History of Scotland (translated) Bishop John Leslie

Even as late as the sixteenth century, the aurora still inspired terror and thoughts of war and bloodshed in its witnesses.

2.3 EXCURSIONS; THE SEVENTEENTH TO EIGHTEENTH CENTURIES

2.3.1 The Maunder Minimum

The mid-seventeenth and early eighteenth centuries saw a period apparently marked by an almost total lack of sunspot activity, now popularly known as the Maunder Minimum (Eddy, 1976). Attention was first drawn to this by Gustav Sporer (1822–95), also noted for his work on the latitude distribution of spots during the sunspot cycle. Later, E. W. Maunder (1851–1928) of the Royal Greenwich Observatory renewed interest in the possibility of the interlude of diminished solar and auroral activity between 1645 and 1715 which now bears his name. Jack Eddy, a solar astronomer at the High Altitude Observatory, Boulder, subsequently brought the idea to major prominence during the 1970s.

Galileo, Scheiner and others began to look regularly at the Sun not long after the telescope's first application to astronomical purposes in 1609, and sunspots were more or less immediately recognized as 'blemishes' on the solar disk. Sunspots were recorded quite regularly until about 1645, following which time they were seen only very infrequently. Flamsteed, writing in 1684, for example, reports:

These appearances, however frequent in the days of Scheiner and Galileo, have been so rare of late that this is the only one I have seen in his face since December 1676.

Chinese astronomers, who made many naked-eye sightings of sunspots in pre-telescopic times, interestingly made few such observations in the period from 1639 to 1720.

At the same time, the records indicate a corresponding dearth of aurorae, from which researchers into long-term solar activity conclude that the shortage of sunspots was genuine, and not merely a result of lack of observer interest. The extensive catalogue of historical auroral sightings (*Verzeichniss Beobachteter Polarlichter*) published by Fritz in 1873, for example, lists only 77 European aurorae during this period, a quarter of them from 1707 to 1708, when sunspot activity may have begun to revive. Many of the reports come from higher latitudes.

The absence of aurorae visible to European observers is a major link in the argument for the occurrence of a period of diminished solar activity during the late seventeenth and early eighteenth centuries. It must be conceded that few aurorae would, indeed, be seen in mid-latitudes were the Sun to cease producing the large spot groups with which are associated the ejections into interplanetary space that trigger them.

The auroral record is but one clue to the past activity of the Sun. Data from other sources, such as the deposition of ^{14}C in annually laid-down tree rings, do indicate that the Sun was less active during a 70-year spell between the seventeenth and eighteenth centuries.

The radioactive isotope ^{14}C is a natural product of the interaction between high-energy (galactic) cosmic rays and nitrogen in the Earth's atmosphere. At times of high solar activity, which are accompanied by a strengthening of the general magnetic field in the solar vicinity, fewer extrasolar cosmic rays can penetrate the inner solar system, and less ^{14}C is therefore produced in the atmosphere.

Atmospheric ^{14}C (from carbon dioxide) becomes assimilated into the newly growing parts of plants. Sections can be taken through long-lived trees, such as the California redwoods, and amounts of ^{14}C incorporated into annual growth rings measured in the laboratory. Results from several independent sources around the world do, indeed, suggest that the late seventeenth to early eighteenth century was a sustained period of high atmospheric ^{14}C abundance, perhaps corresponding with a lull in solar activity.

The period was also marked by low winter temperatures in the northern hemisphere, resulting in the final collapse of the Norwegian colony in Greenland. The frozen River Thames in London was the venue of winter Ice Fairs. It is said of the great English astronomer of the time, Edmond Halley (1656–1743), that one of his most profound desires was to see an aurora. This he eventually did, at the age of 60. Indeed, Halley witnessed and made useful observations of, two splendid aurorae, both passing overhead and into the south of the sky from London, in 1716 and 1719, as sunspot activity began to recover and the familiar 11-year cycle became established.

2.3.2 Other possible excursions

On the basis that the Maunder Minimum may have been only one of several such 'excursions' in solar/auroral activity in historical times, other periods of unusual activity—both high and low—have been sought. Eddy's study of historical solar activity has been tentatively extended further back into pre-telescopic times by examination of the tree-ring ^{14}C record, and of naked eye sunspot observations recorded by Chinese astronomers.

Reasonably reliable records of naked-eye sunspot sightings—made when the Sun's glare was reduced by haze or dust around sunrise or sunset—can be found as far back as 200 BC, the time of the Han Dynasty in China (Clark, 1979; Yau and Stephenson, 1988). Successive Chinese emperors saw the value of favourable celestial portents, and an Astronomical Bureau was established for the purpose of recording them. Caution has to be exercised in any interpretation of these records, however, since there were periods when it was politically advantageous for the portents to go unrecorded.

Despite these limitations, some researchers consider it possible to detect interludes of high and low sunspot activity in the Chinese annals. Evidence is found, for example, of high sunspot activity in the twelfth century period which provided a rich crop of sightings in Scotland and England (section 2.2). Conversely, sunspot activity was apparently low from the late sixth to early ninth centuries, and from AD 1403 to AD 1520.

The ^{14}C record perhaps corroborates these findings. The putative gap in sunspot activity in the fifteenth and sixteenth centuries, for example, coincides with a period of high atmospheric ^{14}C concentration, and has been termed the Sporer Minimum. Low atmospheric ^{14}C concentration, and frequent sightings of both aurorae and naked eye sunspots lend support to the possibility of a Medieval Maximum around the twelfth century.

Several other maxima and minima—some of them more severe in their deviation from 'normal' than the Maunder Minimum—may be postulated as far back as the Bronze Age on the basis of ^{14}C abundance studies, but it is much more difficult to find auroral, sunspot, or other supporting observational evidence.

Eddy has suggested the possibility of long-term cycles of solar activity (spanning as much as 1000 years each), in which excursions such as the proposed Medieval Maximum and Maunder Minimum represent peaks and troughs respectively. It is possible, if Eddy's assumptions are correct, that activity is currently on a gradual rise towards a 'supermaximum' in the twenty-second and twenty-third centuries.

2.4 THE LATE EIGHTEENTH CENTURY

Halley's observations on March 16 1716 and November 10 1719 were followed by a sighting of coronal aurora from France by de Mairan in 1726. This observation was represented on what is believed to be the first 'gnomonic' all-sky summary chart of auroral activity. Scientific interest in the aurora was certainly great during the eighteenth century following the end of the Maunder Minimum, and several records were made of displays over northern Europe at this time.

The naturalist Gilbert White (1720–93) was an assiduous observer of many phenomena from his location in what was then rural England, as published in his *Natural History of Selborne*. White made a number of auroral observations, recorded in his journals between 1768 and 1793 (Tyldesley, 1976). Among his records are such examples as:

> October 25 1769: A vivid aurora borealis, which like a broad belt stretched across the welkin from East to West. This extraordinary phenomenon was seen the same night from Gibraltar.

> January 18 1770: Vast aurora: a red fiery broad belt from E to W.

> February 15 1779: A vivid aurora: a red belt from East to West.

> October 13 1787: The aurora was very red and aweful.

From his location in Hampshire, at latitude 51°N in southern England, White recorded a remarkable number of aurorae. Most were seen in the year or so before sunspot maximum, which seems to be the time at which aurorae are, indeed, commonest at lower latitudes.

2.5 THE NINETEENTH CENTURY

Solar and auroral activity by the beginning of the nineteenth century had settled into the patterns familiar to modern observers. Some very high sunspot maxima occurred, notably in 1836 and 1870. As has proved to be the pattern in the twentieth century, the most vigorous, extensive aurorae tended to be those in the pre-sunspot-maximum phase, an example being a fine display extending to southern England in 1847.

A major event in auroral history was the observation at 11:20 a.m. on September 1 1859 of a solar flare by the English astronomer Richard Carrington. Carrington was surprised to see a brightening within a group of sunspots which he was sketching at the time.

Such 'white light' solar flares are extremely rare: flares are most commonly recorded using equipment which allows the Sun to be observed in the light of hydrogen-alpha, and which had yet to be developed in Carrington's day.

Carrington's flare was followed, a day or so later, by intense auroral activity over much of the Earth. The aurora was even visible in the tropics, from Honolulu, Hawaii (latitude 21°N). Activity persisted for several days. Ground electrical currents associated with the aurora caused disruption of telegraph communication systems in Europe and America.

A brilliant auroral display occurred on October 24 1870. Another huge auroral display—perhaps the most extensive in relatively recent times—occurred on February 4 1872. Aurora was again visible to the tropics on this occasion, with sightings being made from Bombay and elsewhere in India. The display was also seen from such locations as Aden, Egypt, Guatemala, and the West Indies. Associated with this event was another major geomagnetic storm.

The nineteenth century also saw several expeditions to the polar regions, by explorers such as Fridtjof Nansen and Sir John Franklin. Nansen's explorations are recorded in a series of extensive volumes (for example, Nansen, 1897, 1902), in which frequent mention is made of the aurora. Often, as he journeyed northwards away from the auroral zone, Nansen recorded the aurora as confined to the south of his sky. It remains a popular misconception, however, that aurorae should be more common as the observer nears the poles. Nansen's accounts of the aurora are eloquent, to say the least:

Fig. 2.1. The Norwegian explorer Fridtjof Nansen made a number of expeditions to high latitudes, where he frequently saw the aurora. In addition to his eloquent written accounts, Nansen made a number of sketches of auroral forms, meteorological phenomena and wildlife. This fine example shows an auroral band snaking across the sky.

As on the previous night, too, there was a glorious show of northern lights in the southern sky. The great billows of light rolled backwards and forwards in long, undulating streams. The flickering of the rays and their restless chase to and fro suggested crowds of combatants, armed with flaming spears, now retiring and now rushing to the onset, while suddenly as if at given signals huge volleys or missiles were discharged. These flew like a shower of fiery darts, and all were directed to the same point, the centre of the system, which lay near the zenith. The whole display would then be extinguished, though only to begin and follow the same fantastic course again.

(August 7 1889) *The first crossing of Greenland*

Figure 2.2. Another example of Nansen's auroral sketching, from his two-volume travelogue *Farthest North*, showing a multiple-rayed band structure viewed more or less side-on.

Later in the evening Hansen came down to give notice of what really was a remarkable appearance of aurora borealis. The deck was brightly illuminated by it, and reflections of its light played all over the ice. The whole sky was ablaze with it, but it was brightest in the south; high up in that direction glowed waving masses of fire. Later still Hansen came again to say that now it was quite extraordinary. No words can depict the glory that met our eyes. The glowing fire-masses had divided into glistening, many coloured bands, which were writhing and twisting across the sky both in the south and north. The rays sparkled with the purest, most crystalline rainbow colours, chiefly violet-red or carmine and the clearest green. Most frequently the rays of the arch were red at the ends, and changed higher up into sparkling green, which quite at the top turned darker, and went over into blue or violet before disappearing in the blue of the sky; or the rays in one and the same arch might change from clear red to clear green, coming and going as if driven by a storm. It was an endless phantas-magoria of sparkling colour, surpassing anything that one can dream. Sometimes the spectacle reached such a climax that one's breath was taken away; one felt that now something extraordinary must happen—at the very least the sky must fall.

Farthest North

Franklin was a very experienced Arctic explorer, and following his early expeditions brought back several records of the polar aurora. He recognized that the phenomenon could assume a number of forms:

The beams are little conical pencils of light, ranged in parallel lines, with their pointed extremities towards the earth, generally in the direction of the dipping needle.

The flashes seem to be scattered beams approaching nearer to the earth, because they are similarly shaped and infinitely larger. I have called them flashes, because their appearance is sudden and seldom continues long. When the aurora first becomes visible it is formed like a rainbow, the light of which is faint, and the motion of the beams indistinguishable. It is then in the horizon. As it approaches the zenith it resolves itself into beams which, by a quick undulating motion, project themselves into wreaths, afterwards fading away, and again and again brightening without any visible expansion or contraction of matter. Numerous flashes attend in different parts of the sky.

Narrative of a journey to the shores of the polar sea in the years
1819, 1820, 1821, 1822

The British explorer Robert F. Scott, during his expedition to the south pole (which was reached first by Roald Amundsen) early in the twentieth century, also recorded the aurora in his log:

The eastern sky was massed with swaying auroral light, the most vivid and beautiful display that I had ever seen—fold on fold the arches and curtains of vibrating luminosity rose and spread across the sky, to slowly fade and yet again spring to glowing life.

Fig. 2.3. One of the most striking forms of the aurora is the corona, in which the rays and other features appear to converge on a single area of sky as a result of perspective. Such aurora is reasonably common at higher latitudes, and was often noted by Nansen, whose pencil sketch is shown here.

The brighter light seemed to flow, now to mass itself in wreathing folds in one quarter, from which lustrous streamers shot upward, and anon to run in waves through the system of some dimmer figure as if to infuse new life within it.

The appearance of such accounts in various 'travelogues' is no doubt a major reason for the popular perception of the aurora as a phenomenon occurring only in the polar regions.

2.6 AURORAE IN THE TWENTIETH CENTURY

Many outstanding auroral storms have occurred during the twentieth century though, as mentioned earlier, the increasing spread of artificial light pollution has somewhat restricted the visibility of these, particularly in the latter half of the century. Major events in the early twentieth century included those of September 25 1909 (visible from the Cocos Islands, and from Singapore at a latitude of 1°25′N), and May 15 1921 (seen in Samoa at 14°S).

A well-remembered auroral storm was that of January 25 1938, seen widely across a Europe on the brink of war. The display was strikingly red, and was easily visible from Cornwall in southwest England. Sightings were also made from Barcelona in Spain, and Lisbon in Portugal.

The International Geophysical Year in 1957–58 was fortunately timed, coinciding with extremely high sunspot and consequent auroral activity. Displays on September 13 and September 23 1957 were visible from Mexico, as was the 'Great Red Aurora' of February 10 1958. This last event was accompanied by electrical blackouts in several areas of northeastern Canada, as a result of associated ground currents.

The maximum of the following sunspot cycle (cycle 20) was relatively quiet, though a notable series of events did occur in August 1972, again accompanied by geomagnetic effects which caused power fluctuations and communications difficulties in North America. It is fortunate that the violent solar activity responsible for setting off these events did not coincide with any of the Apollo lunar missions: the energetic particles released would have been lethal to astronauts aboard their relatively unshielded vehicle. The effects of this outburst of activity on the interplanetary medium were measured from a number of unmanned spacecraft, including the Pioneers (en route to Jupiter) and by radio astronomers observing distant sources through the disturbed outer solar atmosphere.

Cycle 21 brought several good displays of aurora to mid-latitudes, particularly during 1978–79, and again in 1982. Noteworthy events for observers at lower latitudes included major storms in May and November 1978, and an all-night coronal storm with red rays visible right down to the south of England on March 1–2 1982. Ironically, the most extensive and spectacular display of this entire sunspot cycle occurred almost at its minimum, on the night of February 8–9 1986 (Miles, 1988; Simmons et al, 1990), visible from Hawaii at latitude 20°N.

2.7 THE GREAT AURORA OF 1989 MARCH 13–14

Sunspot cycle 22, which peaked in June 1989, was marked by a very rapid rise in sunspot activity, accompanied by several fine aurorae for observers at lower latitudes.

Of the aurorae during this period, the most dramatic was undoubtedly that seen widely over March 13 and 14 1989. The display was sufficiently bright to be visible even from quite badly light-polluted locations (some observers at dark sites claimed that the aurora itself cast shadows!), and attracted a fair degree of public attention.

The trigger for this storm, as with nearly all aurorae visible to very low latitudes, was a large, complex flare-active sunspot group in the Sun's northern hemisphere. The spot was readily visible to the protected naked eye for several days either side of the auroral storm itself (Verschuur, 1989). Listed as Active Region 5395, the spot group first appeared round the eastern limb of the Sun on March 7: flare activity extending over the limb from the far side of the Sun had already alerted solar astronomers to the possibility that this was an exceptional active area. Several further flares—many of them very violent—were seen in association with the spot group, including one on March 10, which appears to have been associated with the ejection of energetic particles whose arrival at Earth caused the auroral storm.

The aurora, following the observed flare by 36–48 hours, was extensively seen. Initial sightings of auroral activity, coincident with the onset of worldwide magnetic field disturbances, were made from County Clare, Ireland, in the early hours (about 02 hours Universal Time) of March 13 (Gavine, 1989). Activity built through the day, providing a great spectacle for the many around the world who witnessed it.

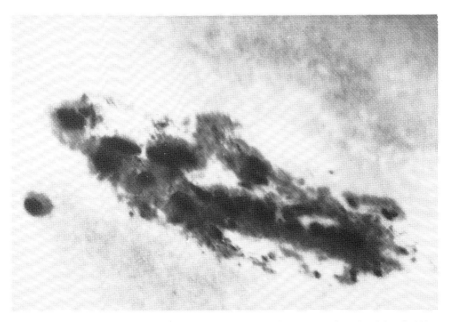

Fig. 2.4. The most spectacular auroral storm seen in recent times occurred on the night of 1989 March 13–14, when activity was seen across much of the civilized world. The activity was triggered by solar flare activity arising in a huge sunspot group, covering 3600 millionths of the visible hemisphere of the Sun. Bruce Hardie photographed this giant sunspot—which was readily visible to the naked eye—on 15 March 1989 at 1051 UT, as it approached the Sun's northwest limb.

Observers in Florida, Arizona, Southern California and elsewhere in the southern United States, were surprised to see blue, green and red auroral forms filling their skies (Eicher, 1989). For many, the most memorable feature was the appearance of vivid red auroral rays: as is evident in records stretching back to medieval times, red colouration is characteristic of the most extensive aurorae. Even Canadian observers, many of them used to regular, fine auroral activity, found this event remarkable.

Sightings were also made from Mexico, and from the Caribbean. The aurora was recorded from Grand Cayman Island at latitude 19°N (Kurth, 1991).

In the antipodes, the aurora australis was visible from New Zealand and Queensland, Australia (latitude 25°S), as reported in the magazine *New Scientist*. The display was also seen from South Africa, at latitudes of 30°S, where the aurora is a rare sight indeed.

European observers also witnessed a fine display—reputedly the best in the Netherlands since the 1938 storm. In the British Isles, skies were clear over most of the country for once, and the aurora was visible from points between the Orkneys in the north, and the Channel Islands in the south. At least 150 detailed reports were received by the Aurora Sections of the British Astronomical Association and Junior Astronomical Society (Livesey, 1989a, 1989b), and the display generated great public interest. Several accounts were received from the general public (Bone, 1989), the following by a Mrs Wylde of Malmesbury, Wiltshire, being typical of how the display was seen by the many for whom this was their first aurora:

The far eastern end of the display began to develop a very strong glow. This bright blob stabilised, no longer pulsating, but growing in size. Suddenly it flashed out several white rays, quite bright and reaching towards the zenith. Then came the colour—all the rays blushed a bright crimson, and for a second or two I saw a rich blue and a trace of glorious golden yellow. The rays, ranging from bright crimson to deepest dull red, came right over the zenith, and seemed to converge. The display grew from horizon to horizon.

This coronal phase, usually short-lived from mid-latitudes, persisted for much of the night for observers in Scotland, for whom the display continued all night, before fading into the dawn. The extreme peak of activity came just before magnetic midnight, around 22 hours local time, and was the phase witnessed by the majority of British observers. At the same time, the aurora was visible to observers in Hungary, Portugal and Spain, and into the Mediterranean area.

Most British national newspapers reported on the aurora. For example, *The Times* covered it under the headline 'Observing the light fantastic'. Elsewhere, *The Guardian* described the aurora as a 'Spring highlight for the south'. Local newspapers also covered the aurora. The Cambridge *Evening News* spoke of a 'Spectacular sight in south for sky searchers'. Aurorae are relatively uncommon in the most densely populated, southermost, parts of the United Kingdom, and it is not surprising that this auroral display was such a source of media interest.

In addition to its visual splendour, this auroral event was accompanied by intense radio effects as the particle influx disrupted the ionosphere high in the atmosphere. Amateur radio operators enjoyed record-breaking communications over long distances—England to Italy, for example, impossible under normal conditions using simple equipment.

The geomagnetic disturbance on March 13–14 was also intense. An index of magnetic activity over a 12-hour period, simultaneously measured from antipodal points in either hemisphere—the *aa index*—suggests the 1989 storm to have been the third largest in the 120 years since records were first made in this way, the most intense having been that of 1925 (section 2.6). The ground currents induced by this activity caused electrical blackouts in Sweden and Canada, as a result of power surges in high-latitude areas. In Quebec Province, six million people were left without electricity for up to nine hours—testimony, indeed, that geomagnetic storms can have important effects on human affairs.

Visual observers at higher latitudes saw activity continue into the next night, though at gradually diminishing levels. Within a couple of days of the storm, all was quiet. Some waited with great anticipation for the return of Active Region 5395 one solar rotation later, and the possibility of a recurrence of the intense aurora in early April 1989. By then, however, the sunspot had decayed, and nothing out of the ordinary was seen in terms of aurora.

2.8 OTHER EVENTS IN CYCLE 22

As discussed later (section 7.2.4), there appear to be two peaks of auroral activity at lower latitudes in most sunspot cycles. The Great Storm of 1989 March 13–14 was, without doubt, the most spectacular event in the first peak associated with sunspot cycle 22. The display had been preceded, for those at higher latitudes at least, by good auroral activity throughout the autumn of 1988.

Following the Great Storm by a matter of weeks was a further extensive display on 1989 April 25–26, which was also visible to the latitudes of southern England. Thereafter, auroral activity at lower latitudes dwindled for a time: with the exception of a faint aurora on July 28–29, briefly visible to the latitudes of southern England, 1990 was a very quiet year for aurorae at lower latitudes, despite continued high sunspot numbers.

This comparative lull was brought to an end by a vigorous auroral display, seen extensively in the United States on 1991 March 23–24 (Eicher, 1991a), and continuing into the following night, when it was visible right down to the South Coast of England. From Scottish latitudes, this aurora, with interludes of brilliant red coloration, was coronal for much of the night. The display was triggered by flare activity associated with the long-lived Active Region 6555, the second-largest (covering 3000 millionths of a solar hemisphere) spot group of cycle 22, which had crossed the Sun's central meridian a few days previously.

Another extensive (2500 millionths of a solar hemisphere) spot group, Active Region 6659, crossed the Sun's disk on 1991 June 9, and was reportedly even more flare-active than that responsible for the Great Aurora in 1989 March. Forecasts, widely picked up on by the news media, suggested that aurora would be visible to very low latitudes as a consequence. In the end, the anticipated events around June 12 amounted to comparatively little, except for those in the United States (Eicher, 1991b) who had darker skies than their colleagues at the latitudes of the British Isles. Activity was extensive for only a short time.

The expectation of good auroral activity as 1991 drew to a close was borne out by an event visible to northern Italy on October 1–2, and the exceptional November 8–9 display (section 2.8.1). Late December of 1991 was also graced by some fine, if less extensive events, the outbursts of activity ending with another short-lived display, this time over western European longitudes, on 1992 February 1–2.

2.8.1 The Major Aurora of 1991 November 8–9

That the aurora can sometimes spring surprises is perhaps best demonstrated by the outstanding event of 1991 November 8–9. Activity during the preceding months had admittedly been high, with several fine displays extending to lower latitudes (section 2.8). During the first week of November, however, the Sun appeared relatively quiet, with the white-light projected image showing only a few, relatively small, spot groups. Despite this comparative dearth of sunspots, it became apparent, as soon as darkness fell, that a major aurora was in progress. Like the Great Aurora of March 1989, this event was witnessed by many thousands of amateur astronomers and others around the world (Bone, 1992, McEwan, 1992 and Eicher, 1992).

Intense auroral activity was visible from early evening onwards, even at the latitudes of the English Midlands. A group of beginners, gathered for an astronomy course at the Preston Montford Field Centre near Shrewsbury was treated to spectacular views of brilliant red auroral rays filling the northern sky. As the night went on, the aurora intensified, and even along the South Coast of England, the aurora stretched to the zenith at times. The ray-tops were visible from the south of France.

In southwest Scotland, experienced observer Tom McEwan recorded coronal activity lasting all night, with an auroral band running through Orion's Belt to the *south* of the sky. Quite simply, this was a stunning display, perhaps even more intensely coloured than the March 1989 storm, and a good target for photographers. A great many pictures were taken by amateur astronomers, showing the dominant red rays and green bands.

Activity continued unabated as night fell over the United States some hours later. The aurora was visible down to the southern states of Texas, Georgia, Alabama and Oklahoma.

The major aurora of 1991 November 8–9 came at a particularly poignant time for magnetospheric physicists at the University of Iowa, four of whose colleagues (Chris Goertz, Dwight Nicholson, Bob Smith and Yinhua Shan) were shot dead the previous week by a disaffected graduate student. Memorial services for the four were held close to the date of the aurora; scientists at Iowa regard the aurora as having been an appropriate natural tribute in itself.

No obvious candidate sunspot group appeared to be in the right place on the solar disk to set off this major auroral storm. Rather, it is believed that a disturbance—perhaps a flare in a spot group on the Sun's averted hemisphere—produced a shock-wave, which in turn caused detachment of a *filament* (equivalent to a solar prominence seen in profile against the bright disk). A candidate filament disappeared from the Sun's SE quadrant on November 6. Material from the filament, thrown into the solar wind, arrived at Earth around 48 hours later, resulting in the magnetospheric disturbance and spectacular auroral display recorded on November 8–9.

While it is true, in general, that major aurorae most often follow the central meridian passage of large, active sunspot groups across the solar disk, reliably forecasting the occurrence of low-latitude auroral activity remains difficult. Great Aurorae such as those of 1909, 1925, 1938 and 1989 are usually associated with the sunspot cycle's rise to maximum. As evidenced by extensive events near sunspot minimum in 1986, however, the aurora will doubtless continue to surprise observers for a long time to come.

REFERENCES

Bone, N. (1989) The Great Auroral Storm of 1989 March 13–14. *Astronomy Now* **3** (6) 22–29.

Bone, N. (1992) Autumn aurora provides spectacle over UK. *Astronomy Now* **6** (1) 8–9.

Clark, D. (1979) Our inconstant Sun. *New Scientist* **81** (1128) 168–170.

Dicks, D. R. (1970) *Early Greek Astronomy to Aristotle*. Thames & Hudson.

Eddy, J. A. (1976) The Maunder Minimum. *Science* **192** 1189–1202.

Eicher, D. J. (1989) Brilliant Aurorae produced by solar flare. *Astronomy* **17** (7) 95–97.

Eicher, D. J. (1991a) Night of the aurora. *Astronomy* **19** (8) 77.

Eicher, D. J. (1991b) The glowing sky of June. *Astronomy* **19** (10) 91.

Eicher, D. J. (1992) Night of the Great Aurora. *Astronomy* **20** (3) 89–91.

Gavine, D. (1989) *The Astronomer* **25** (300) 247–248.

Kinder, A. J. (1990) The progress of astronomy in England—Earliest times to 1558. *J. Brit. Astron. Assoc.* **100** (4) 182–190.

Kronk, G. W. (1988) *Meteor showers: a descriptive catalogue*. Enslow.

Kurth, W. S. (1991) The great solar storms of 1989. *Nature* **353** 705–706.

Livesey, R. J. (1989a) BAA Aurora Section Newsletter 15.

Livesey, R. (1989b) The Great Aurora of 1989 March 13/14. *J. Brit. Astron. Assoc.* **99** (3) 113–114.

McEwan, T. (1992) The greatest show on Earth. *Popular Astronomy* **39** (2) 15–16.

Miles, H. (1988) Magnetic storm and auroral display of 1986 February 8–9. *J. Brit. Astron. Assoc.* **98** (4) 194–199.

Nansen, F. (1897) *Farthest North: being the record of a voyage of exploration of the ship Fram 1893–96.* Constable.

Nansen, F. (1902) *The first crossing of Greenland.* Longmans, Green.

Simmons, D. A. R., Henriksen, K., Taylor, M. J., and Hermansen, D. (1990) A remarkable outburst of solar activity and its geomagnetic effects. *J. Brit. Astron. Assoc.* **100** (6) 280–286.

Tyldesley, J. B. (1976) Gilbert White and the Aurora. *J. Brit. Astron. Assoc.* **86** (3) 214–218.

Verschuur, G. (1989) The day the sun cut loose. *Astronomy* **17** (8) 48–51.

Yau, K. K. C., and Stephenson, F. R. (1988) A revised catalogue of far eastern observations of sunspots (165 BC to AD 1918). *Q. Jl. R. astr. Soc.* **29** 175–197.

Zhang, Z. (1985) Korean auroral records of the period AD 1507–1747 and the SAR arc. *J. Brit. Astron. Assoc.* **95** (5) 205–210.

3

Scientific investigations of the aurora

3.1 EARLY IDEAS

Early observers and theorists of the aurora classed it along with other atmospheric phenomena as a 'meteor'. In common with many of his other ideas which remained unchallenged until well into the sixteenth century, Aristotle's view of these events being the result of ignition of rising vapours below the innermost celestial sphere must have prevailed for some time. An alternative, proposed by the Roman philosopher Seneca in his *Questiones Naturales*, was that aurorae were flames viewed through chasmata—cracks in the heavenly firmament.

In the twelfth-century Norse chronicle *The King's Mirror*, three possible causes are given for the aurora. It is described as either a fire encircling the Earth; creation of light by the Sun; or light absorbed and re-emitted by ice and snow.

The great sixteenth-century Danish naked-eye observer Tycho Brahe (1546–1601) noted a similarity between aurorae and the haloes produced around the Moon by thin clouds. The French mathematician and philosopher Rene Descartes (1596–1650) considered aurorae to originate from sunlight reflected by particles in cirrus clouds.

3.2 THE SEVENTEENTH CENTURY

A major aurora was observed on September 12 1621 by the French astronomer Pierre Gassendi (and also, apparently, by Galileo in Venice). Following this event, Gassendi introduced the term *aurora borealis* (from the Latin 'northern dawn'—an apt description of the aurora's appearance from lower latitudes) to describe the phenomenon (Chapman, 1967). It has also been suggested, however, that Gregory of Tours may have used the name aurora borealis as early as the sixth century AD (section 1.4).

Meanwhile, scientific understanding in other, related areas was growing apace. Terrestrial magnetism, long exploited by seafarers for navigational purposes, was studied by William Gilbert (1544–1603) and others. Gilbert arrived at the conclusion that the Earth could be envisioned as a giant dipole magnet. Later, in 1716, the English astronomer Edmond Halley came close to the suggestion that auroral rays were the result of particles flowing along magnetic field lines. Terrestrial magnetism was a major area of interest for Halley, who took a series of measurements of the magnetic dip in the north Atlantic.

3.3 THE EIGHTEENTH CENTURY

Halley had to wait until late in life to see his first aurora, having lived through the Maunder Minimum period of apparently diminished sunspot and auroral activity (section 2.3.1). His first display, on March 16 1716 was sufficiently extensive to appear overhead for a time from London:

> Out of what seemed a dusky Cloud, in the N.E. parts of the Heaven and scarce ten Degrees high, the Edges whereof were ringed with a reddish Yellow like as if the Moon had been hid behind it, there rose very long luminous Rays or Streaks perpendicular to the Horizon, some of which seem'd nearly to ascend to the Zenith. Presently after, that reddish Cloud was propagated along the Northern Horizon, into the N.W. and still further westerly: and immediately sent forth its Rays after the same manner from all Parts, now here, now there, they observing no Rule or Order in their rising. Many of these Rays seeming to concur near the Zenith, formed there a Corona ... .
>
> (From Halley's 'Account of the late surprising appearance of the Lights seen in the air on March 16 last', *Philosophical Transactions of the Royal Society* **29** (1716).)

A further, similarly extensive display was observed by Halley on November 10 1719. Halley measured the position of the central point of the corona (a perspective effect in which auroral rays appear to converge when overhead) on this second occasion.

A major student of the aurora at this time was the Frenchman Jean-Jacques d'Ortour de Mairan (1678–1771). de Mairan is credited with the first measurements of auroral height, made in 1726. He also noticed the seasonal effect in the frequency with which aurorae are detected at lower latitudes, which peaks around the equinoxes. His theories included the suggestion that the solar atmosphere, extending all the way to the Earth, was involved in generating the aurora. A catalogue of auroral records was produced by de Mairan, published in his book *Traite Physique et Historique de l'Aurore Boreale* under the auspices of the French Academy of Sciences in 1733.

Around this time, George Graham (1674–1751) in London began taking detailed measurements of the daily fluctuations in the local magnetic field. This work was extended in 1741 by the much-travelled Swedish physicist, astronomer and mathematician Anders Celsius (1701–1744) and his assistant Olof Hiorter, who discovered a correlation between days of disturbed magnetic field and auroral activity. Celsius and Graham found that magnetically disturbed days in Uppsala, Sweden, coincided with days which were also disturbed in London.

The first European to observe the aurora australis was Captain James Cook, on February 17 1773, while in the Indian Ocean near latitude 58°S. The display was recorded in his log:

> ... between midnight and three o'clock in the morning, lights were seen in the heavens, similar to those in the in the northern hemisphere, known by the name of Aurora Borealis ... .

His observations confirmed the suggestion of de Mairan that there should be aurora around the southern, as around the northern, polar regions.

The latter parts of the eighteenth century saw further theories put forward to explain the aurora, including that by Benjamin Franklin, presented again through the French Academy of Sciences. Franklin suggested that aurorae were produced by a 'lightning' effect as mobile hot air from the tropics descended from great heights into the high polar atmosphere.

3.4 THE NINETEENTH CENTURY

The nineteenth century was a time of rapid improvement in understanding of the aurora, paralleled by advances in other, related areas. Around 1844, the existence of the roughly 11-year sunspot cycle was discovered by Heinrich Schwabe, a German pharmacist and amateur astronomer who carefully monitored the Sun on every possible occasion in the hope of finding the supposed intra-Mercurial planet Vulcan.

The recognition of a solar-auroral connection still lay some way off, however, and even as late as the 1890s there was learned speculation as to whether sunspots were influenced by the Earth's action on the Sun, rather than vice versa (BAA Memoir, 1948). Some workers, long pre-dating the now-discounted 'Jupiter effect' (which failed to generate the predicted earthquakes and other havoc in the early 1980s) suggested that the orbital positions of Jupiter and Saturn influenced the length and nature of the sunspot cycle.

Strong hints to the association between solar activity and enhanced auroral activity must surely have been given following Carrington's observation of a white-light solar flare and the extensive auroral activity on subsequent nights (section 2.5).

During the early nineteenth century, it became recognized that aurorae occurred most frequently within certain latitudinal zones, as surmised by the German geographer Muncke in 1833 and the Yale professor Elias Loomis in 1860. Many catalogues of auroral records were compiled by nineteenth-century researchers, including one by Hermann Fritz in Zurich. Fritz introduced the term *isochasms* to describe geographical places sharing the same frequency of auroral visibility.

Spectrographic techniques were applied to auroral observation by the Swedish physicist Anders Jonas Ångstrom in 1867, who found the aurora to produce a strong emission line in the yellow-green region of the spectrum. The wavelength of this emission—the 557.7 nm 'auroral green line' resulting from excitation of atomic oxygen—was not precisely measured until much later. Identification of its nature was made by McLennan and Shrum at Toronto in 1924.

3.5 THE TWENTIETH CENTURY

Attempts had been made in the eighteenth and nineteenth centuries to measure the heights of aurorae using the technique of visual triangulation, which was successfully applied to the study of meteors by the German observers Brandes and Benzenberg in 1798. The most prolific series of auroral triangulation observations was obtained by the Norwegian Carl Stormer and his colleagues in a photographic programme initiated in 1911.

The aurora is a pleasing, if sometimes awkward, photographic subject. Improvements in emulsion speeds and technology are certainly an advantage for modern observers

(Henderson, 1990), who endure none of the difficulties encountered by the Danish meteorologist Sophus Tromholt, to whom the first successful attempt at auroral photography, in 1885, is attributed (Petrie, 1963).

Photographic emulsions had improved sufficiently by Stormer's time that the images of auroral features obtained by his Norwegian observing network could readily be used for measurements and triangulation. Stormer and his colleagues worked from about 20 observing stations, each separated by at least 20 km, and in telephone contact with each other so that exposures could be made simultaneously. Observing in often extremely cold conditions, they used robust Krogness cameras, which contained a minimum of moving parts that could become frozen or jammed.

Basically, the Krogness camera consisted of a dark chamber to hold a 10×14 cm ('half-plate') glass plate, coated with photographic emulsion, and a fast (usually $f/1.5$) cine lens which could be slid to six positions: each plate therefore contained six different exposures. Exposures were made by flipping up a light-tight flap over the lens. The camera swivelled on an altazimuth mount and had a simple cross-wire sight.

One of Stormer's original Krogness cameras is in the possession of the British Astronomical Association Aurora Section.

Fig. 3.1. David Gavine (Assistant Director of the BAA Aurora Section) with one of the robust Krogness cameras of the type used by Stormer and his colleagues to obtain multiple-station images of the aurora for triangulation. The camera's simplicity was ideal for work in freezing conditions. As shown, exposures were made by flipping up a lens cover. Six images could be made on a single plate, which was slid into fixed positions behind the lens.

Observations by Stormer's network did much to reveal the distribution of activity around the auroral oval. Some 40 000 photographs were taken during this work, enabling over 12 000 accurate auroral measurements to be made.

3.5.1 Other advances in the twentieth century

During the early twentieth century, important developments in methods for observing the Sun were also made. George Ellery Hale and his colleagues at Mt Wilson Observatory in America devised the spectrohelioscope allowing observation of the Sun in single, isolated wavelengths, in the late nineteenth century. Routine observations of the Sun in the wavelengths of hydrogen-alpha or ionized calcium became possible. Activity in the chromosphere (the inner atmosphere immediately above the bright surface of the photosphere) was opened to observation, bringing the discovery that this region was far from quiescent! The correlation between flares visible in the wavelength of hydrogen-alpha, in the sunspot regions and the occurrence of aurorae at mid-latitudes became accepted.

Further extension of spectroheliographic equipment allowed detection of the magnetism associated with sunspots by virtue of the Zeeman effect (in which spectral lines

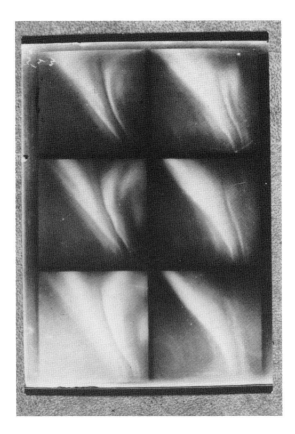

Fig. 3.2. A typical Krogness camera plate, with six recorded images of the aurora.

become split under the influence of a nearby strong magnetic field) in 1908. Using such equipment, the reversal of the magnetic polarities in sunspot regions from one sunspot cycle to the next was found.

The French astronomer Bernard Lyot used monochromatic filters for solar observation in the wavelength of hydrogen-alpha light from 1930 onwards. Around this time, Lyot also developed his coronagraph, an optical means of producing an 'artificial eclipse' in the telescope, allowing high-altitude observatories to monitor activity in the inner solar corona on a routine basis. Other new technologies were emerging, allowing more detailed investigation of the region of the Earth's atmosphere in which the aurora occurs. Radio, developed for communication purposes, was first used to transmit signals across the Atlantic from Britain to America by Marconi in 1901. Marconi was able to send radio waves over this vast distance thanks to their reflection from an ionized layer in the upper atmosphere. This *Heaviside layer*, corresponding to what is now more commonly termed the ionospheric E-region, lies at about 110 km altitude. It comprises a layer of increased electron density, resulting from the action of short wavelength solar radiation on the upper atmosphere.

Later investigators, from the 1920s onwards, used radio waves as a means of probing the high atmosphere for further layers of increased electron density. The F layer, between 160–300 km altitude, was found by Appleton during studies of the ionosphere in 1924. The processes by which ionization occurs in the high atmosphere were opened to study during the 1930s. Laboratory studies on the mechanisms of gas ionization in discharge tubes led to improved understanding of the processes of recombination and attachment of electrons to produce negative ions. The means by which certain of the auroral 'forbidden' emissions are produced also became clearer from such work.

In 1940, Harang and Stoffregen made the first investigations of radio wave reflection from auroral ionization in the high atmosphere. Amateur radio operators in the United States also carried out much work of this nature. This period also saw the first detection of radio waves from the Sun (and, indeed, the birth of radio astronomy in general), allowing solar activity to be followed over a still wider range of electromagnetic wavelengths.

3.5.2 The development of auroral theory to the 1950s

By the beginning of the twentieth century, the scene was set for major developments in theories to explain the cause of the aurora. Around 1880, the electron was discovered. Studies showed that the movement of electrons could be influenced by both electrical and magnetic fields. Extending these findings, the Norwegian physicist Kristian Birkeland arrived at the suggestion, in 1896, that aurorae are most frequently seen in the polar regions, since it is towards here that electrons from an external source are directed by the Earth's magnetic field, and that the Sun might be a source of of streams of such fast-moving electrons.

An enthusiastic observer from northern Norway (like his contemporary Stormer, he attempted to obtain auroral height measurements), Birkeland also conducted laboratory experiments to expand his theories. A magnetized iron ball (called a *terella* by Birkeland), painted with fluorescent material, was suspended in a vacuum chamber and bombarded with electrons ('cathode rays'), and the distribution of the induced fluorescence determined. The results led Birkeland to develop his ideas of how electron beams could stimulate high-latitude aurorae.

Stormer pursued this idea from a theoretical mathematical, rather than experimental, approach, and arrived at conclusions broadly similar to those of Birkeland.

The recognition that solar activity in some way influenced terrestrial aurorae, starting with the finding that solar flares were frequently followed by displays reaching lower latitudes, was reinforced by the findings of Julius Bartels. Extending a finding made as long ago as 1856 by the Scottish physicist Broun, Bartels found that periods of geomagnetic disturbance often recurred at 27-day intervals. Broun had, initially, considered this to be a lunar effect, but it was later realized to more likely be tied to the apparent rotation of the Sun as seen from Earth. Bartels described these stream-producing centres on the Sun as 'M-regions' (magnetic or mystery regions).

While M-region particle streams could account for active periods, there was as yet no complete acceptance of the possibility of continuous outflow from the Sun during quiet intervals. The velocities necessary to carry particles from the Sun to the Earth to produce auroral storms—up to 1500 km s^{-1}—were, however, appreciated.

Important developments in the theories to account for geomagnetic storms were made by Sydney Chapman and Vincenzo Ferraro in the 1930s. In particular, their work led to the development of ideas regarding the region in which Earth's magnetic influence is dominant—a volume of space named the *magnetosphere* by Thomas Gold some 20 years later. Chapman and Ferraro pictured this as producing a cavity in ionized plasma streams emerging from the Sun. The terrestrial magnetic field was subject to compression and confinement within the cavity.

From further observation, it became apparent that the solar atmosphere as a whole must be continually expanding into interplanetary space, including the immediate environs of the Earth. Ludwig Biermann introduced the idea that the plasma tails of comets might provide indirect means of observing variations in the outflow from the Sun. A fuller model for this solar wind was proposed by Eugene Parker in 1957, and subsequently confirmed by early spacecraft missions (Parker, 1964).

Theoretical developments thereafter centred on clarifying the interactions between the solar wind and the magnetosphere which result in the production of auroral activity.

3.6 THE POLAR YEARS

The aurora is a phenomenon which occurs on a global scale. An observer at a single location sees only a small fraction of the whole pattern. It became apparent that groups of observers taking standardized simultaneous measurements from a number of locations could usefully pool their results. Collaborative international ventures on the grand scale have become an important feature of scientific research during the twentieth century. The International Polar Years provided early examples of how such research could be carried out.

The first Polar Year, in 1882–83 was aimed at obtaining information not only on the aurora, but also on a wide range of meteorological and other phenomena in the Arctic regions. Eleven nations participated.

Fourteen countries took part in the Second International Polar Year (SPY) 50 years later, during 1932–33. Observations were made to a standard scheme detailed in an auroral atlas and supplement, and which remained valid at least 20 years later (International Geodetic and Geophysical Association/Union, 1932, 1951).

Many more geomagnetic stations were set up within the Arctic Circle. As on the previous occasion, a major Canadian–British station was established at Fort Rae, the reports from which have been described as classic examples of how these should be prepared (Chapman, 1969). Edward Appleton and his colleagues set up a radio sounding station at Tromso in Norway.

3.7 PERMANENT AURORAL OBSERVATORIES AT HIGH LATITUDES

It has been stressed by many commentators that a great deal of research carried out during the Polar Years was from temporary observatories, often with rather primitive equipment. Valuable work was done, however, in the years following SPY from permanent auroral observatories at high latitudes. These included several in North America, such as Gartlein's observatory at Ithaca, NY, Saskatoon, Saskatchewan, and College, Alaska. European observatories included those at Tromso and Oslo in Norway, and Kiruna in Sweden.

The relatively dependable occurrence of auroral activity at these locations allowed more intensive study with advanced equipment. Fast 'all-sky' cameras could be used to study the distribution and forms of the aurora through the course of the night. Detailed spectroscopy using diffraction gratings was carried out from observatories such as Yerkes. Radar observations allowed continuing expansion of earlier studies on the ionospheric regions in which the aurora appears.

Sounding rockets allowed direct investigations of atmospheric conditions at high altitudes during aurorae. Black Brant rockets, capable of reaching altitudes up to about 900 km, have been launched for auroral study from the Fort Churchill Range in Canada, the Poker Flats Range in Alaska, and elsewhere. Launches have also been made from Kiruna, and other European locations.

Much of this more advanced scientific equipment was in place in time for the International Geophysical Year of 1957–58, which was to yield a great deal of useful information relating to auroral phenomena and their causes.

3.8 THE INTERNATIONAL GEOPHYSICAL YEAR

The International Geophysical Year (IGY), which actually ran for the 18 months between July 1 1957 and December 31 1958, saw international collaboration in auroral and atmospheric study on a scale still greater than the SPY. Detailed planning of the scientific programme to be pursued began in 1952, under the umbrella of the International Council of Scientific Unions. This included continuous visual, photographic and radio monitoring of solar activity and study of its subsequent effects—including auroral activity—on the terrestrial environment. Meteorological and oceanographic observations also had a role in the overall collection of data for the IGY: basically, the aim was to gather as much information on a wide range of inter-related phenomena from as many geographical locations as possible (Martin, 1957).

Auroral studies formed an important part of the IGY effort. The value of visual observations was recognized, and amateur astronomers in Britain, the United States and elsewhere were actively encouraged to participate by taking measurements and providing descriptive accounts of any auroral activity seen from their locations. Reports of this nature, again made following standard guidelines, were also collected from meteorological observers, sailors and aircrew, considerably extending the geographical range covered.

Professional observers operated a network of all-sky cameras around the polar regions in order to take simultaneous images of the aurora's distribution. Spectrographic analyses were also carried out, and radar techniques applied to the study of ionospheric movements during auroral activity.

Balloons and rockets were used for high altitude cosmic ray measurements. Sounding rockets—among them the British Skylark, developed just in time for the IGY—were launched into regions of the atmosphere affected by auroral displays to obtain *in situ* measurements of particle energies and densities. Skylark was capable of reaching heights up to about 210 km, well into the auroral layer of the atmosphere above 100 km altitude. America had ambitious plans to launch several artificial satellites into Earth orbit to supplement other IGY activities.

The problems encountered by the US Vanguard programme have been well documented (Wilson, 1985), and the IGY saw Russia take the initial lead in the 'space-race' with the launch of Sputnik 1 on October 1 1957. Sputnik was equipped only with a radio transmitter. The first successful American launch, however, the 14 kg Explorer 1 satellite which entered Earth orbit on January 31 1958, was significant in carrying instruments capable of measuring its local radiation environment. Using this equipment, regions containing trapped energetic particles girdling the Earth—the Van Allen belts—were discovered. The detailed exploration of the near-Earth space environment had begun.

The Vanguard programme eventually placed a satellite in orbit on March 17 1958. Like Sputnik 1, this carried only a transmitter, but tracking of Vanguard 1 was of great use in geodetic studies. Over the following six months, two more Vanguard and two more Explorer satellites were launched.

An important aspect of the IGY was the extension of the observational network to the southern polar region; the Polar Years had involved research solely in the Arctic. The Royal Society's research base at Halley Bay in Antarctica was established in time for workers there to participate in IGY studies.

Like so many other scientific ventures, IGY did not yield its full fruits until some years after the observations—an immense collection of data—had been made. An important finding, resulting from analysis of all-sky photographic observations, was the conclusion by Feldstein and Koroshove in 1963 that the aurora is distributed in oval regions around either geomagnetic pole (Akasofu, 1979). Understanding the mechanisms which control the behaviour of these auroral ovals has been important in subsequent and continuing attempts to unravel the workings of solar–terrestrial interactions.

The IGY was timed to coincide with the maximum of sunspot cycle 19, which turned out to be the highest on record. This was a time when great auroral displays were seen, reaching to mid-latitudes. Many of the observers hold fond memories of this period, when amateur and professional astronomers were able, together, to contribute to scientific advances.

3.9 THE SPACE AGE

The success of the IGY was followed by the International Years of the Quiet Sun (IQSY) in 1964–65, intended to study the same phenomena under the conditions prevailing at sunspot minimum. By then, satellites had become more commonly and reliably available for studies of the highest parts of the atmosphere *in situ*, and from above. Results from early American Explorer spacecraft had, by this time, begun to unravel the structure of the magnetosphere (Cahill, 1965).

Spacecraft exploration of the solar system has led to a greater understanding of the environments around the major planets. The interplanetary medium has also been investigated. The Mariner 2 mission to Venus in 1962 (Wilson, 1987) carried magnetometer equipment, from which confirmation of the solar wind predicted by Parker was obtained. Subsequent investigations of the interplanetary medium have been made by the Interplanetary Monitoring Platform (IMP) series of probes. These have yielded valuable information about the solar wind and the complicated processes which go on within it. Similar measurements have since been taken in the distant outer reaches of the solar system by the Pioneer and Voyager spacecraft.

Auroral studies continue from an extensive network of ground-based observatories. The equipment available grows ever more sophisticated, but observations are still supplemented by photographic and, to a lesser degree, visual means. Low-light television cameras allow recording of the often rapid motions of auroral forms at high latitudes; some of the spectacular results have been made available as commercial videos for home consumption.

Many observing stations for auroral and other research now operate in the Antarctic. The Japanese-funded Syowa station is well placed for auroral observations. Halley Base, part of the British Antarctic Survey, is ideally located for studies of auroral effects at the boundary of the plasmasphere (a magnetospheric region in which particles are trapped). Radio investigations, including work with the Advanced Ionospheric Sounder (AIS) and assessment of ionospheric particle densities using a riometer (*r*elative *i*onospheric *o*pacity *meter*) are carried out from Halley (Rycroft, 1985).

Sounding rockets still provide the means of briefly sampling conditions in the auroral layer. Facilities such as EISCAT also allow more detailed study of ionospheric processes during aurorae. EISCAT (European Incoherent Scatter facility) is a collaborative venture between several European countries. UHF radio signals broadcast from a 32-metre dish at Tromso in Norway, and reflected from auroral structures, can be detected by dishes at Kiruna (Sweden) and Sodankyla (Finland), and at Tromso itself. VHF signals are also transmitted from, and received back at, a 120 by 40 metre 'trough' at Tromso. Measurements allow ionospheric particle densities and movements to be assessed (Williams, 1985). EISCAT is being extended by the addition of facilities at Spitsbergen. A further addition to the network of HF radars is the Co-operative UK Twin Located Auroral Sounding System (CUTLASS) at stations in Finland and Iceland, which commenced operation in 1995. Funded by the UK, Finland and Sweden, CUTLASS is designed to monitor ionospheric particle movements.

EISCAT receivers have also been used as conventional radio telescopes for detection of the scintillation of distant cosmological sources (quasars), produced as a result of large-scale movements of plasma in the solar wind. Similar work was carried out at Cambridge from 1964 to 1981 (Hewish, 1988), during which the first pulsar was fortuitously discovered. Hewish and his colleagues were able to detect 'squalls'—potential sources of magnetic disturbances—moving out through the solar wind.

NASA aircraft flights in 1967, during which simultaneous photographs of auroral forms visible at conjugate points in the northern and southern hemispheres were taken, fulfilled the expectation that activity in one auroral oval should mirror that in the other. Polar-orbiting spacecraft in the US Defense Meteorological Satellite Program (DMSP) have been used to image the auroral ovals from above. It is said that, when US military satellites first recorded activity of the northern auroral oval over Siberia, this was interpreted as a sophisticated form of Soviet camouflage!

Only small sections of the ovals may be seen from low Earth orbit, an unsatisfactory situation for those wishing to study their overall behaviour. The problem was solved by the launch of satellites with highly elliptical polar orbits, capable of looking down on an entire hemisphere of the Earth around apogee. The Canadian ISIS-II satellite, launched in 1970, was the first to provide such complete images.

The most staggering images of the auroral ovals in their entirety must surely be those returned from the NASA Dynamics Explorer-1, launched in August 1981, as one of a pair of satellites (Reddy, 1983). The eccentric orbit of Dynamics Explorer-1, with apogee at 22 000 km (3.5 Earth-radii) allowed the whole auroral oval to be visualized for intervals of up to 5 hours. The time-series of images recording the progress of a substorm, presented in Colour Plate 2, provides an example of the advantages of such coverage.

The sister-satellite Dynamics Explorer-2 was placed in a lower, more circular orbit, and provided complementary particle and other measurements, before re-entering Earth's atmosphere in February 1983.

Following the IMP series, further exploration of the near-Earth space environment—both in the magnetosphere and the solar wind—has been carried out by a number of spacecraft, including the Russian Prognoz satellites. Of particular note was ISEE-3 (International Sun–Earth Explorer-3), launched as part of a multi-satellite mission in 1977 to an initial stable position 'upwind' of Earth at the L1 Lagrangian libration point between the Earth and the Sun. Later orbital manoeuvres sent it—renamed ICE (International Comet Explorer)—to the first near encounter between a spacecraft and a comet (Giacobini-Zinner) in 1985. A veritable flotilla of spaceprobes from Europe, Japan and the Soviet Union later encountered Comet Halley in March 1986, further adding to knowledge of plasma interactions in the solar wind. The European Giotto probe, its camera blinded by a dust impact during its passage close to Halley's nucleus, was subsequently successfully targeted to a second comet (Grigg-Skjellerup), obtaining further valuable measurements in July 1992 (section 5.4.1).

Ion releases from satellites have been used to further map the magnetosphere. A successor to ISEE was another multi-satellite mission, AMPTE (Active Magnetosphere Particle Tracer Explorer), jointly funded by the US, UK, and Germany. Launched in 1984, this carried out a number of barium and lithium releases, whose attempted detection from component satellites at various locations in the magnetosphere allowed improved models to be derived for particle movements. Just after Christmas 1984, a release from AMPTE produced an 'artificial comet' in the solar wind, whose development was followed by a number of instruments, including some at ground-based observing stations (Coates and Smith, 1985).

3.10 SPACECRAFT IMPROVE KNOWLEDGE OF SOLAR PHENOMENA

Solar astronomy—a crucial link in unravelling the auroral mechanism—has also benefited greatly from advances in satellite technology. Ground-based observatories such as those at Kitt Peak, Big Bear, and La Palma, are eminently capable of routine coverage of the Sun's activity in many electromagnetic wavelengths. Observations at X-ray and ultraviolet wavelengths, however, had been restricted to fleeting glimpses from rockets making brief flights above most of the obscuring atmosphere.

Satellites provided the opportunity to make longer-duration observations, unimpeded by atmospheric absorption, at the wavelengths strongly emitted by energetic solar phenomena. A series of Orbiting Solar Observatories (OSO 1-8) was launched by NASA, followed by the highly successful manned Skylab space-station in 1973–74 (Furniss, 1986).

Skylab included a number of instruments for solar observation, notably the Apollo Telescope Mount, with which images at ultraviolet and X-ray wavelengths were obtained. The Skylab missions, three in all lasting a total of 171 days, coincided with the declining phase of solar cycle 20, which reached its minimum in 1976. Skylab observations confirmed the existence of coronal holes—persistent regions of reduced solar magnetic field and atmospheric density. These appear particularly during the later years of the sunspot cycle, and in many respects can be regarded as corresponding to Bartels' M-regions: the steady, long-duration particle streams in the solar wind, injected into the solar wind via coronal holes can cause geomagnetic disturbances, recurring at 27-day intervals. Two persistent coronal holes were observed in detail from Skylab, including the 'Boot of Italy' feature, so named for its profile against the solar disk.

Skylab also revealed the existence of X-ray 'hot-spots' (apparently coincident with the bases of long streamers in the solar corona, and of short-lived X-ray bright features, which may be a major source of turbulence in the solar wind (Parker, 1975).

These observations were followed up by another successful satellite, the Solar Maximum Mission (SMM, or 'Solar Max'), launched in February 1980 (Ryan, 1981). As its name suggests, SMM was aimed primarily to study the Sun at a different phase of the sunspot cycle from that observed during the Skylab missions. After a few months of virtually perfect operation, however, many of the detectors aboard SMM were disabled by an electrical fault in April 1981. Repairs were carried out in orbit in 1984, in time for SMM to cover the *minimum* between cycles 21 and 22 during 1986 in great detail, before the increased atmospheric density resulting from intense solar flare activity early in cycle 22 brought about its demise.

SMM obtained detailed observations of solar flare and other phenomena in X-ray and ultraviolet wavelengths. Coronagraph observations from SMM resulted in the discovery of several comets, as did the Solwind instrument aboard a US Military satellite—P78-1—whose fate was sealed by the Star Wars Programme, in the development of which it was later used for target practice!

3.11 SPACE-AGE THEORETICAL PROGRESS

The availability, from the early 1960s onwards, of satellites and spacecraft with which to investigate the orbital and interplanetary environments has allowed more detailed development of theories to explain how the aurora is generated, and why it behaves as it does. Soon after results from the early Explorer and IMP spacecraft were obtained, James W. Dungey began the development of models wherein the aurora was driven by reconnection processes between interplanetary and magnetospheric magnetic field lines. This work has been extended by Akasofu and many others, providing good working explanations for why some solar disturbances give rise to major auroral activity, while others do not.

From a study of simultaneously taken photographs by ground observers around the auroral ovals, obtained during the IGY, Syun-Ichi Akasofu in the 1960s developed the substorm model for auroral behaviour at high latitudes. It became recognized that these events are distinct from the more extensive geomagnetic storms which carry the aurora to lower latitudes. Both substorm and geomagnetic storm aurorae have been imaged from above by Dynamics Explorer and other satellites, confirming the IGY-based deductions, and allowing further refinement of theory.

Progress in understanding the events underlying the solar phenomena which, propagated via the solar wind, drive the auroral mechanism, has also been great following the advent of satellite monitoring of the Sun. Fine-scale modelling of solar flares and their causes has been made possible by the use of equipment capable of providing the necessary extremely detailed observations.

The solar–terrestrial link is becoming ever better understood through the use of scientific tools, of which our forebears can scarcely have dreamed.

3.12 THE FUTURE OF AURORAL AND SOLAR RESEARCH

While an improved picture of the complex interactions between the Earth's magnetic field and that of the solar wind in which it is imbedded has been gathered over the years, much remains to be elucidated, and the aurora is still an object of concentrated study. Exploration of the magnetosphere and solar wind will continue, with the Cluster and SOHO missions, due for launch in 1995–96. The CRRES (Combined Release and Radiation Effects) satellite, launched in 1990, began, in January 1991, a series of gas releases in the near-Earth environment aimed to follow up the earlier work of AMPTE and ISEE.

SOHO (Solar and Heliospheric Observer) is scheduled for launch from the NASA space shuttle in late 1995. It will operate outside the magnetosphere, around the L1 libration point of the Earth's orbit, there gathering information about the solar atmosphere and solar wind. SOHO instruments will also monitor solar oscillations (section 4.3.1). Observations will be controlled from the Goddard Space Flight Centre. Cluster will be launched by an ESA Ariane rocket, and will comprise a group of four satellites, taking simultaneous measurements of plasma movements and interactions in the magnetosphere from different locations. The Cluster satellites are each equipped with four 50-metre and two 5-metre instrument-tipped booms. Together, Cluster and SOHO represent a collaboration between ESA and NASA, involving scientists in France, Germany, Switzerland, Sweden, the United Kingdom and the United States.

Fig. 3.3. The SOHO satelliite, launched in late 1995, will carry out studies of solar activity which affects the Earth from some 1.5 million km 'upwind' at the L1 Lagrangian point. ESA photograph.

SOHO and Cluster, along with earlier missions such as Geotail, Intersat, Wind and Polar, are part of a wider cooperative effort, the International Solar Terrestrial Physics Programme (ISTP), involving a wide range of observations from many locations. For example, British Antarctic Survey scientists are collaborating with those in Greenland (at equivalent geomagnetic locations in opposite hemispheres) in a series of radar observations. The interplanetary scintillation measurements made from Cambridge in the 1960s, 1970s and early 1980s will be extended in future years using larger instruments.

The Skylab and SMM studies of the Sun have been followed in the early 1990s by the immensely successful Yohkoh (Japanese for 'Sunbeam') orbital solar observatory. A joint Japanese, British and American venture, Yohkoh was launched on a planned 3-year mission in August 1991, carrying instruments for the detection of low- and high-energy (soft and hard) X-rays, and gamma rays (Petersen *et al*. 1993). Imaging of energetic processes in the solar corona at these wavelengths has been supplemented by ground-based observations.

Prior to the ESA Ulysses mission, launched from the Space Shuttle *Discovery* in October 1990 (Moore, 1991), observations of phenomena in the solar atmosphere have, of necessity, been confined to those occurring near the ecliptic plane. Following a 'slingshot' encounter to within 450 000 km of Jupiter in February 1992, Ulysses was flung into a

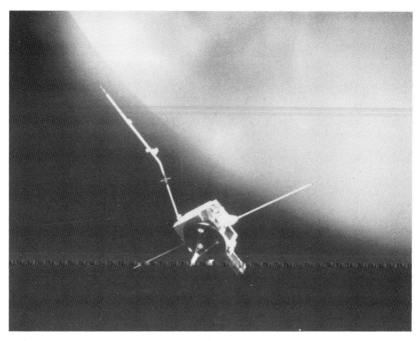

Fig. 3.4. Since its launch in 1990, the Ulysses probe has been hugely successful in unravelling the nature and behaviour of the solar wind and magnetic field outwith the ecliptic plane. ESA photograph.

high-inclination orbit around the Sun, at a slightly greater distance than that of the Earth. Ulysses made a pass over the southern pole of the Sun from late June to early November 1994, reaching a maximum ecliptic latitude of 80.2°S on 1994 September 13. A north polar pass followed a year later.

Ulysses carries instruments for measurement of solar wind ion composition, and with which to examine the interactions between cosmic rays and the solar wind. Magnetometers aboard Ulysses will study the interplanetary magnetic field, and radio and plasma wave phenomena are also measured. It is planned to extend the Ulysses mission into the early twenty-first century, affording an opportunity to compare conditions at solar minimum and solar maximum.

REFERENCES

Akasofu, S.-I. (1979) Aurora borealis. *Alaska Geographic* **6** (2).

BAA (1948) British Astronomical Association: the first fifty years. *BAA Memoirs* **42**, Part 1 (1948).

Cahill, L. J. (1965) The magnetosphere. *Scient. Am.* **212** (3) 58–68.

Chapman, S. (1967) History of aurora and airglow. In: McCormac, B. M. (Ed.), *Aurora and airglow*. Reinhold.

Chapman, S. (1969) Auroral science, 1600 to 1965. Towards its Golden Age? In: McCormac, B. M., and Omholt, A. (Eds.), *Atmospheric Emissions*. Van Nostrand Reinhold.

Coates, A., and Smith, M. (1985) The Earth's magnetic field yields its secrets. *New Scientist* **108** (1486) 32–34.

Furniss, T. (1986) *Manned spaceflight log.* Jane's.

Henderson, J. (1990) Northern lights. *Photography* **51** 57–60.

Hewish, A. (1988) The interplanetary weather forecast. *New Scientist* **118** (1613) 46–50.

International Geodetic and Geophysical Association (1932) *Supplements to the photographic atlas of auroral forms.*

International Geodetic and Geophysical Union (1951) *Photographic atlas of auroral forms.*

Martin, D. C. (1957) The International Geophysical Year 1957–8. *Penguin Science News* **45** 7–28.

Moore, P. (1991) To the poles of the Sun. *Astronomy Now* **5** (1) 16–20.

Parker, E. N. (1964) The Solar Wind. *Scient. Am.* **210** (4) 66–76.

Parker, E. N. (1975) The Sun. In: *The Solar System.* W. H. Freeman.

Petersen Collins, C., Bruner, M., Acton, L., and Ogawara, Y. (1993) Yohkoh and the mysterious solar flares. Sky and Telescope **86** (3) 20–25.

Petrie, W. (1963) *Keoeeit—the story of the aurora borealis.* Pergamon.

Reddy, F. (1983) Celestial winds, polar lights. *Astronomy* **12** (8) 6–15.

Ryan, J. (1981) A new look at the nearest star: the Solar Maximum Mission. *Astronomy* **9** (5) 6–16.

Rycroft, M. (1985) A view of the upper atmosphere from Antarctica. New Scientist **108** (1484) 44–49.

Williams, P. (1985) European radar unscrambles the ionosphere. *New Scientist* **108** (1485) 46–52.

Wilson, A. (1985) *Frontiers of Space.* Hamlyn.

Wilson, A. (1987) *Solar System Log.* Jane's.

4

The active Sun

4.1 THE IMPORTANCE OF THE SUN TO TERRESTRIAL LIFE

The aurora is only one manifestation of the numerous and complicated interactions between the Sun and the Earth. All life on Earth is ultimately dependent on the energy of the Sun. Photosynthesis by green plants, using the energy of sunlight, is the main source of Earth's atmospheric oxygen. Ultraviolet light from the Sun is harmful to living organisms, but is substantially prevented from reaching ground level by the stratospheric ozone layer—itself a product of the action of solar ultraviolet on atmospheric oxygen. It is solar energy which drives the weather systems, maintaining atmospheric circulation in a relative equilibrium so that the Earth as a whole neither bakes nor freezes. Ancient solar energy, locked up in fossil fuels, is currently exploited to power transport and other essential human requirements.

The very appearance of life, believed to have occurred in the first billion years after the Earth's formation (Horgan, 1991), must surely have been dependent on the Sun's relative stability from an early age. Had the Sun been significantly more massive, it would have expended its nuclear fuel long before life could have become established. Conversely, had the Sun been much less massive, it would in its youth have produced less of the ultraviolet radiation believed by many theorists to have been essential in driving prebiotic chemical processes. The continued long-term stability of the Sun, for which evidence may be found in the sedimentary record (Giovanelli, 1984; Bracewell, 1988), has also been of obvious importance in the development of life on Earth.

Many ancient cultures perceived the Sun as the epitome of perfection. The view of a flawless solar orb persisted in western civilization, unchallenged on religious and other grounds, from the tradition of the Greek philosopher Aristotle in the third century BC until the time of Galileo. Throughout the centuries, however, the meticulous astronomical observers of the Far East had occasionally noted the blemishes of sunspots on the solar disk viewed with the naked eye through dense haze near sunset (section 2.3.2).

Soon after the first application of the telescope to astronomical observation around 1610, it was recognized that the Sun's face was indeed less than perfect, and the existence of sunspots was quickly accepted. Some historians of astronomy suggest that one of Galileo's major crimes in the eyes of the Vatican authorities was not so much in making public the existence of sunspots, but rather in claiming their discovery as his own, when they had previously been noted by Christoph Scheiner, a Jesuit priest. The story that Galileo destroyed the sight of one eye by recklessly observing the Sun directly through his telescope would appear to be apocryphal.

While the Sun is a relatively stable star, it does have certain detectable degrees of variability in its activity. The most obvious of these is the roughly 11-year cycle in which sunspot numbers rapidly reach a maximum then gradually dwindle away. This activity is routinely monitored by a number of professional observatories around the world, and may also be readily followed with simple equipment by amateur astronomers using the safe technique of projection to view the solar disk (Baxter, 1972).

Variations in the total energy output of the Sun—the so-called *Solar Constant*—have been measured in detail by orbiting satellites, notably the very successful Solar Maximum Mission. Measurements suggest that the Sun's energy output is 0.1% greater at sunspot maximum than at sunspot minimum, as a result of the increased emissions from sunspot regions and their associated faculae (Foukal, 1990). Such variations could in turn give rise to a global temperature change of the order of 0.1 K: the contribution of human industrial activities to global warming is, however, far more significant.

There are indications that past solar variability could have had a greater influence on global climate. During the Maunder Minimum of the late seventeenth to mid-eighteenth centuries (section 2.3.1), sunspots appear to have been virtually absent and global temperatures dropped by 0.5 K, perhaps corresponding with a 0.2–0.5% decrease in the Solar Constant.

These changes apart, the Sun is still certainly a great deal more stable than many of the variable stars which may be observed in the night sky. Direct imaging of stellar disks is impossible over the vast distances involved, but astronomers can infer the existence of starspots covering large fractions of the surfaces of some stars. Other stars are prone to flares many times more violent than those which emerge from the Sun during periods of high sunspot activity.

Observations at wavelengths of ionized calcium have been used to monitor chromospheric magnetic activity, which is correlated to brightness variations, in stars similar to the Sun (Lockwood *et al.* 1992). It has been suggested from calcium-wavelength observations that some stars may spend as much as one-third of their time in a 'Maunder Minimum' low-activity state (Baliunas and Saar, 1992).

It is to the events which accompany high sunspot activity that we must look for the root cause of the most spectacular terrestrial aurorae. Even at times of low sunspot numbers, however, other forms of solar activity may still disturb the Earth's magnetic field and generate increased auroral activity.

4.2 THE ORIGIN OF THE SUN

The Sun is only one of some 200 billion stars, grouped together in the Galaxy, which itself is only one of uncounted billions in the Universe. On any clear night, stars at many different stages of their development may be observed in the solar neighbourhood of the Galaxy. These range from young, hot blue stars such as those in the Pleiades cluster, to cool red giants nearing the end of their lifespan, such as Betelgeuse in Orion. Clouds of gas and dust in the Milky Way are believed to be the sites of star formation, while other nebulous objects—supernova remnants such as the Crab Nebula—mark the demise of massive stars which have run through their life cycle.

From observation of the stars, it has become possible for astrophysicists to propose mechanisms for the formation and subsequent evolution of stars of various masses. Such studies suggest the Sun to be a typical cool, yellow dwarf star, in its fairly stable middle age. Its diameter of 1 400 000 km, while enormous by terrestrial standards, is far from exceptional for such a star.

Best current theories suggest that the Sun, and its planetary retinue, originated from the gravitational collapse of part of a cloud of gas and dust—perhaps similar to the Orion Nebula—some 5×10^9 years ago. A trigger has to be sought for the initiation of the nebula's contraction. Candidates include passage of the nebula through the pressure wave associated with one of the Galaxy's spiral arms, or the arrival of shock waves from a supernova explosion in the nebula's vicinity. The latter hypothesis may be supported by observed isotopic abundances in meteorites, some of which are believed to represent residual primitive material which failed to become incorporated into larger solar system bodies.

As the pre-solar nebula began to contract, dust particles within the cloud clumped together into progressively larger aggregates. Most of the material aggregated together in the centre of the contracting cloudlet to form the Sun, while smaller collections came together to form the planets and their satellites. Some of the condensed material was not swept up by the larger bodies, however, and remains in the form of meteorites, asteroids and comets. Further details of the solar system's origin may be elucidated by direct spacecraft examination of asteroids and comets.

A critical phase in the development of the Sun came when its mass grew sufficiently large that the exterior pressure raised core temperatures to the level where nuclear reactions were initiated: the Sun began to shine.

Since that point, the Sun has been in a fine balance between the inwards pressure of gravity from its outer layers, and the outwards radiative pressure from its core. This balance will be maintained for the next 5×10^9 years, until the hydrogen which undergoes nuclear fusion in the Sun's core runs out. At that stage, further nuclear reactions based on the fusion of helium will take over, and the Sun will expand enormously in size to become a red giant, engulfing the Earth. Eventually, this secondary source of solar energy will run out, and the Sun will shrink to become a white dwarf star, perhaps surrounded for a time by a planetary nebula.

Stars more massive than the Sun are more intrinsically brilliant, but run through their nuclear fuel more rapidly, and have much shorter lifetimes. The blue giants Rigel and Vega, for example, will have lifetimes of only a few millions of years, before ending in dramatic supernova explosions as their radiative and gravitational pressures depart from equilibrium in the course of a few seconds (Murdin, 1990).

4.3 ACTIVITY ON THE SUN

4.3.1. The solar furnace

The innermost third of the Sun comprises its superdense, extremely high temperature core, a region within which nuclear reactions occur continuously. Physical models of the Sun's structure suggest core temperatures of the order of 15 million kelvins. Under these conditions, fusion of hydrogen to generate helium can occur. These nuclear reactions, predominantly the proton–proton reaction, are the source of solar light and heat. Photons generated during these reactions emerge after long intervals following their 'random walk' outwards through the overlying layers. It has been estimated that, due to collisions with particles in the overlying radiative zone, each photon produced in the core of the Sun takes some 10^4–10^5 years to finally emerge at the surface (Nicolson, 1982).

Another physically predicted by-product of the proton–proton reaction in the Sun's core are neutrinos. Unlike photons, these have negligible interactions with matter, and emerge from the solar surface virtually unimpeded after their creation. Attempts have been made to detect solar neutrinos by measuring their rare interactions with large volumes of appropriate chemicals. Collisions between neutrinos and atoms of chlorine can give rise to measurable amounts of a radioactive isotope of argon, ^{37}Ar. Accurate measurements of the levels of ^{37}Ar generated in a large tank of chlorine-containing cleaning fluid, 1500 metres underground in a former gold mine at Homestake in South Dakota, suggest that the solar neutrino flux is only a third of that predicted. This apparent dearth of solar neutrinos—the so-called neutrino problem—is a source of some concern to astrophysicists, and remains an area of active investigation.

A genuine dearth of solar neutrinos might require a revision of models of the solar interior, placing constraints on conditions in the core. For example, the observed neutrino capture rate could be accounted for by a lower core temperature. Such solar models, however, also require that some mechanism exist to allow the core to be supported against the inwards pressure of the overlying layers. This might be achieved if the core were rapidly rotating, or if abundances of heavy elements such as iron were sufficiently different from those expected that convection could occur at core depths. Alternatively, core temperatures may vary over time, such that currently observed solar radiation levels are consistent with a past higher core temperature, while the observed neutrinos, produced by core reactions today, indicate a lower temperature at the present epoch: perhaps the radiative flux will drop in the far future when the more sluggish photons finish their random outwards walk.

The field of *helioseismology*, in which the subtle oscillations of the Sun's outer layers are studied, allows some further investigation of both the surface and the solar interior (New, 1990). The results of this work do not suggest existing models of the solar interior to be in error (Leibacher *et al.* 1990).

Several other explanations for the neutrino problem have been forwarded, including suggestions that existing detectors may not be sufficiently sensitive, or that the types of neutrinos emerging from the Sun are such that only one in three may, indeed, be detected. Further investigations by European and Russian physicists, using more sensitive gallium-based neutrino detectors, may help to resolve the problem.

4.3.2 Radiative and convective processes

Overlying the Sun's core is the *radiative zone*, in which little bulk motion is expected. The outer third, the *convective zone*, extends from the deep solar interior up to the visible surface, the photosphere.

Convection on several scales is apparent in the Sun. Most obvious are the small 'rice-grain' convective cells, observable under good conditions through even quite small Earth-

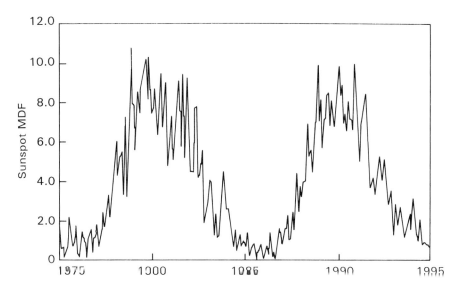

Fig. 4.1. The most obvious indicators of solar activity are sunspots, whose numbers may be followed on a daily basis using even quite simple equipment. The mean daily frequency (MDF) of sunspot groups is calculated monthly. The plot here, based on amateur observations, covers cycles 21 and 22. The rapid rise to maximum of cycle 21 between minimum in 1975–76 and its peak in 1979–80 is clearly seen, along with the subsequent more gradual decline to sunspot minimum in 1986. The rise of Cycle 22 was extremely rapid, leading some authorities to suggest that it would be the highest observed in telescopic times: in the event, the peak (reached in June 1989) was the third highest recorded. Activity remained close to maximum levels for an extended period, before declining towards minimum expected in 1996 The *average* interval between maxima is around 11 years, but it remains difficult to predict the Sun's exact behaviour.

based telescopes as granulation in the brightly shining photosphere. Small convection cells are on the order of 1000 km to a side, and have fairly short lifespans—of the order of 10 minutes. Larger convection cells in the photosphere—the supergranulation—tend to be longer-lived, persisting for perhaps 24 hours. The boundaries of supergranules, which may be 30 000 km to a side, are frequently marked by spicules and other indicators of intense magnetic field activities.

4.3.3 Sunspots and the solar magnetic field

The outermost layer of the Sun does not rotate as a solid body, showing, instead, a *differential rotation*. The relative motions of sunspot groups at different solar latitudes suggest a rotation period of about 25 days for the solar equator, and as long as 36 days for the poles. Helioseismology shows differential rotation to affect the convective layer to a depth of 200 000 km. Below this level, the rotation does seem to be continuous. The strong magnetic field of the Sun appears to be a consequence of this differential rotation, coupled with convective processes.

Convection in the outer third of the Sun may amplify the solar magnetic field and bring it to the surface. Movements of the highly conductive solar gas are determined by the magnetic field: gas is constrained to flow along, but not across, solar magnetic field lines. As a result, the smooth pattern of the convective granules can be disrupted in regions

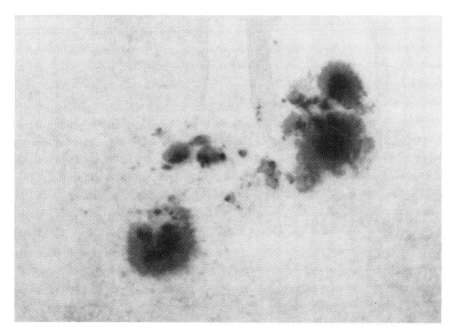

Fig. 4.2. A typical, well-developed sunspot group photographed on 1988 September 2 by Bruce Hardie (Director of the BAA Solar Section). Leader and follower spots are seen, with smaller activity regions between them. The dark central umbrae and lighter penumbral regions are obvious, and hints can be seen of the granulation which results from convection in the photosphere.

where magnetic flux emerges from the solar interior through the photosphere. Gas rising to the surface in such regions is restrained by magnetic fields whose strength may be of the order of 1000–3000 Gauss (for comparison, the Earth's magnetic field has an intensity of 0.3–0.6 gauss). Prevented from moving laterally and sinking again, the gas cools, resulting in the appearance of sunspots (Noyes, 1990).

Sunspots appear dark against the surrounding, hotter photosphere. Typical sunspot temperatures are of the order of 4000 K, compared with 6000 K for their immediate surroundings. They are therefore not particularly cool by terrestrial standards: if a large sunspot could be isolated from the photosphere and suspended in the night sky, it would provide much more illumination than the full Moon! The photosphere itself is fairly shallow in comparison with the convective layer as a whole. The visible light of the Sun comes from the top 100 km or so, levels much below this being opaque to light.

Sunspots show wide variations of size, appearance, structure and duration. The smallest sunspots are simple pores, lasting only a few hours. More complex groups may begin life as pores, growing in size to affect large areas of the solar disk. Large sunspots show two distinct regions, a darker (and cooler) central umbra, surrounded by a lighter penumbra. Structures in the penumbra show alignment to strong local magnetic fields. Complex spot groups may contain several umbrae embedded in a mass of penumbra, and surrounded by pores. The appearance of such groups, which can become sufficiently large to be visible to the protected naked eye, may change from hour to hour: such rapidly changing, complex spot groups are often the source of solar flares.

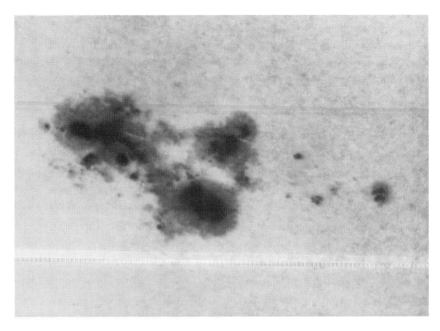

Fig. 4.3. A large, complex sunspot group, seen early in the rise of Cycle 22. Photographed by Bruce Hardie on 1988 June 30, this group was sufficiently large to be visible to the naked eye, using suitable precautions.

The largest sunspot groups may come to cover quite significant areas of the solar disk. Apparent sunspot areas may be estimated using calibrated projection disks. Extensive groups, such as that involved in producing the Great Aurora in March 1989, may cover as much as 3600 millionths (0.4%) of a solar hemisphere (Strach, 1989). By comparison, the Earth is of insignificant size.

The magnetic fields associated with sunspots were first observed by Hale and his colleagues at Mt Wilson in 1908, who found that leader and follower spots within a group show opposite magnetic polarities. Magnetic polarities are reversed, however, for leader and follower spots in the other hemisphere of the Sun at the same time. The picture is further complicated by a reversal of leader and follower spot magnetic polarities in either hemisphere at the end of each approximate 11-year cycle: the complete magnetic cycle takes about 22 years to return to its starting configuration.

The most widely accepted model for the production of sunspots is that proposed by H. W. Babcock in the early 1960s, and subsequently developed by R. B. Leighton. This model accounts both for the appearance of sunspots, and the reversal of the solar magnetic field which accompanies each new sunspot cycle.

Solar magnetic field lines may be initially visualized as lying along meridians running north–south. Differential rotation in the outer layers soon begins to stretch field lines horizontally, concentrating them. Where several field lines are brought into close proximity loops of magnetic flux are forced to the surface, breaking through the photosphere where they disrupt convection and give rise to spot groups.

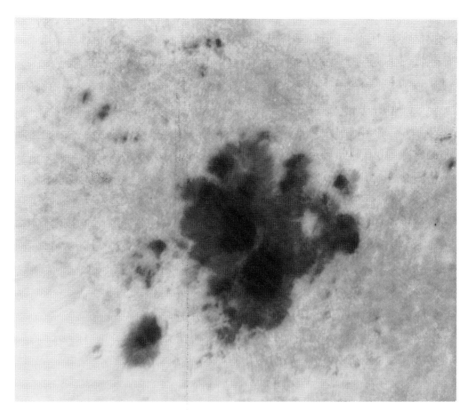

Fig. 4.4. A large sunspot group photographed by Bruce Hardie on 1989 August 12 at 0928 UT.

Field loops initially break through at higher heliographic latitudes. The first spots of a cycle therefore emerge at higher solar latitudes. The zones of latitude at which spots appear most frequently gradually migrate equatorwards. Plotted against time, the distribution of spot latitudes gives rise to the Maunder 'butterfly diagram'. Cycles overlap, so that the first, high latitude, spots of a new 11-year cycle begin to appear while the last near-equatorial spots of the previous cycle are present.

The reversal of polarity in sunspot groups from one cycle to the next is accounted for by the gradual dispersal of magnetic flux from decaying active regions. Leader and follower spots appear at slightly different latitudes (a consequence of the winding-up of magnetic field lines by differential rotation), such that the leader is usually closer to the equator. As an active region decays and its magnetic flux diffuses outwards, more of the leader's magnetic flux will cross the equator into the other solar hemisphere. Consequently, towards the end of one cycle and the beginning of the next, each solar hemisphere comes to acquire a net magnetic polarity derived from that of the leader spots in the opposite hemisphere during the cycle immediately past. When each new cycle starts, magnetic polarities therefore begin from a reversed configuration.

Associated with the regions in which sunspots appear are *faculae*, observed in white light as brighter areas of the photosphere. Faculae are seen to best advantage when close to the limb of the projected disk, where they appear brighter in contrast against the limb darkening (a consequence of the absorption of light from the photosphere by its passage through a greater volume of solar atmosphere in line of sight to the observer than at the centre of the disk). They persist for some months, often appearing before sunspot activity

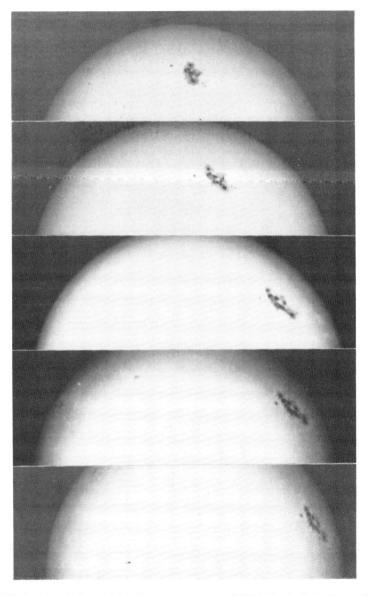

Fig. 4.5. A series of views of the giant sunspot group of 1989 March. Solar flare activity associated with this group gave rise to the Great Aurora of 1989 March. Subtle changes in structure can be seen as the active region (designated AR 5395) crosses the disk. Faculae are also visible at the Sun's eastern and western limbs. The series of photographs, again by Bruce Hardie, was obtained on 1989 March 11, 13, 15, 16 and 17.

manifests itself at that position on the disk, and remaining for some time after the associated sunspots have decayed.

Faculae lie in the upper reaches of the photosphere, corresponding roughly with the plages observed in hydrogen-alpha wavelengths in the overlying chromosphere (section 4.3.4). These appear to be regions of enhanced gas density and temperature.

The active regions in which sunspots are involved are important in generating aurorae. The most vigorous aurorae are generally those which follow the ejection, from the neighbourhood of sunspot groups, of energetic particles during solar flares. The large, actively changing sunspots with which such flares are most frequently associated are commonest in the run-up to sunspot maximum, and this, too, is the time when aurorae are most often seen in mid-latitudes.

4.3.4 Activities observed at hydrogen-alpha and X-ray wavelengths: the chromosphere and corona

The photosphere is the most accessible region of the Sun for white light observation, and sunspots are the most obvious manifestation of magnetically disturbed conditions near the solar surface. Sunspots, however, are only part of the extensive and complex combination of effects which comprise active regions on the Sun. Activity in the chromosphere and corona overlying sunspot groups is of great importance with respect to solar–terrestrial relations.

The *chromosphere*, 10 000 km deep and lying immediately above the photosphere, is briefly visible during total solar eclipses as a ring of red light surrounding the dark body of the Moon. Its colour is produced by excited hydrogen, emitting at a wavelength of 656.3 nm. Filters and spectrohelioscopes have been developed which allow astronomers to study the Sun in this restricted wavelength, the hydrogen-alpha line. Such observations have revealed much about activity in the chromosphere.

Chromospheric temperatures are higher than those of the photosphere, of the order of 10 000 K. A sharp transition in temperature is seen between the two layers. Heating of the chromosphere probably results from upward-travelling shock waves from the photosphere, and also downwards transfer of heat from the inner corona.

When present, solar prominences extending a short distance outwards from the Sun's limb may be a striking feature during total eclipses. These consist of relatively cool gas from the chromosphere, or condensed from the corona above, suspended in the inner solar atmosphere by magnetic fields. The gas in prominences remains cool as a result of the insulating effect of the associated magnetic fields: conduction of heat from the surrounding, hotter corona is inhibited.

The use of hydrogen-alpha filters allows prominence activity to be routinely monitored on a day-to-day basis. Prominences are seen to best advantage when presented on the limb, reaching heights of as much as 50 000 km above the photosphere. They may also be visualized when transiting the disk as *filaments* in hydrogen-alpha, appearing dark in contrast against the brighter background of the Sun. Prominences are relatively long-lived features, and may persist for up to four or five months.

Prominences occasionally erupt from the chromosphere outwards into the corona. These events are associated with coronal transients, or mass ejections. The eruption of a prominence in transit may be seen by Earth-based observers as a 'disappearing filament'.

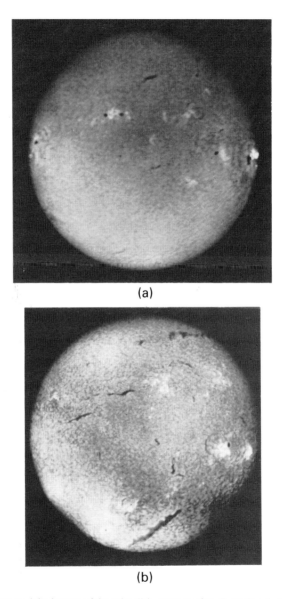

(a)

(b)

Fig. 4.6. Hydrogen-alpha images of the solar disk courtesy of Dr E. H. Strach. (a) Such images reveal activity in the chromosphere lying immediately above the photosphere. Sunspot regions are seen as bright *plages*. A limb flare is in progress in this image. (b) Solar prominences, consisting of gas suspended in magnetic fields above the chromosphere (as seen in Fig. 4.7), appear as dark *filaments* by contrast.

Such events sometimes unload particles Earthward into the solar wind (section 4.3.8), and may be followed a day or so later by enhanced geomagnetic activity and aurorae at lower latitudes. In two-thirds of cases, the prominence becomes re-established within a couple of days of a disappearance event.

When observed as filaments against the solar disk, prominences are seen to lie along the boundary between areas of opposite magnetic polarity in active regions, the *neutral line*. Hydrogen-alpha spectroheliograms show active regions outlined as bright areas—*plages*—against the solar disk.

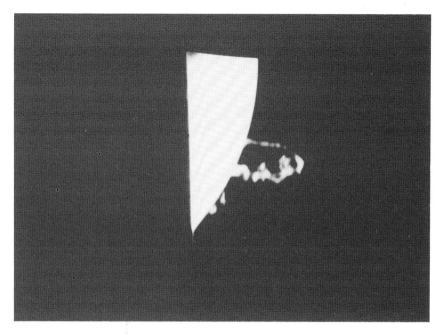

Fig. 4.7. A view of the solar limb in hydrogen-alpha light, showing a prominence extending above the chromosphere into the inner corona. The loop structure is characteristic of the magnetic fields above sunspot regions. Courtesy of Commander H. R. Hatfield.

Prominence activity is often, but not always, associated with sunspot regions. Prominences often appear in decaying active regions once the sunspots have broken up. The frequency with which quiescent prominences are seen at equatorial and tropical heliographic latitudes roughly follows the sunspot cycle, peaking a couple of years later than the spot groups. Active prominences, produced by condensation of material falling back from the corona following violent ejections, are seen at times of high solar activity.

In addition to revealing prominences, total solar eclipses also afford glimpses of the *corona*, the extended region of the solar atmosphere overlying the photosphere and chromosphere. For a long time, the corona was held to be a relatively static, unchanging part of the Sun, altering its shape in response to the sunspot cycle but doing little else. At sunspot maximum, the corona is fairly evenly distributed around the Sun, while at sunspot minimum, it appears drawn out into long equatorial streamers, and is absent from the Sun's polar regions.

The differing configuration of the corona from sunspot maximum to sunspot minimum reflects the changing nature of solar magnetic fields. At sunspot maximum, the tenuous coronal gases are largely confined by closed loops of magnetic fields (having both ends of

Fig. 4.8. Total solar eclipses afford a brief glimpse of the inner solar atmosphere as the pearly corona, surrounding the dark body of the Moon. The corona's structure is dominated and constrained by magnetic fields. Closed field loops above active regions produce bright 'helmets' in the inner corona, while streamers emerge from those regions where the field becomes stretched at great distances from the Sun. Open regions in the coronal magnetic field are believed to be the source of the solar wind, which flows out from the Sun to permeate the solar system. Photograph of the 1968 total solar eclipse, with sunspot activity close to maximum, courtesy of Patrick Moore.

their field lines embedded in the Sun) above active regions. By sunspot minimum, 'open' magnetic fields with one end embedded in the Sun and the other extending outwards to interstellar space come to predominate, and coronal gas becomes spread out more readily.

Two principal types of structure can be observed in the white-light corona: streamers and 'helmet' structures. The bright helmets are produced by coronal gas trapped under the closed magnetic field loop above an active region. Streamers, on the other hand, emerge more or less radially from regions where solar magnetic field lines are open. Long streamers are thus more frequently present and obvious in the corona around sunspot minimum.

This established view of the corona's behaviour with respect to the sunspot cycle may need to be revised in the light of results obtained using specialized photographic techniques during total solar eclipses (Akasofu, 1994). The streamers so prominent at sunspot minimum are thought to delineate the equatorial plane of a simple dipole. Imaging of the outer corona, however, does reveal the presence, even at sunspot maximum, of long faint coronal streamers in the equatorial plane of the 'equivalent dipole' beyond 2.5 solar radii. The equivalent dipole is the sum of a central dipole aligned with the solar rotation axis, and a large-scale equatorial dipole produced by magnetic structures near the Sun's surface. The brightness of the complex inner corona at solar maximum results in such outer structures often being overlooked. The equivalent dipole turns over as the solar cycle progresses, being aligned to the rotational (heliographic) equator at sunspot minimum, roughly perpendicular at maximum, then aligned to the equator again, but turned by 180°, at the following minimum.

The inner parts of the Sun's corona can be routinely studied from high-altitude ground-based observatories by use of the Lyot coronagraph, which provides, optically, an artificial eclipse within the telescope. Ground observers, however, still obtain their most detailed views of the outer corona during rare total solar eclipses. On such occasions, scattering by Earth's atmosphere of sunlight from outside the zone of totality still results in a fairly bright sky background against which it is difficult to resolve fine-contrast features in the outer corona. The best observations of the corona are thus obtained using coronagraphs aboard orbiting satellites.

Satellite observatories have the advantage of being able to exploit emissions in the far ultraviolet and X-ray wavelengths, which allow the corona to be traced to even greater distances from the Sun. X-ray wavelength observations made during the Skylab missions in 1973 and 1974 greatly advanced understanding of solar coronal processes, and laid the foundations for subsequent studies with the SMM and Yohkoh satellites.

The corona is a strong emitter of X-rays as a consequence of its high temperature. Gas in the corona is at yet higher temperatures than that in the chromosphere below, typically $1–5 \times 10^6$ K. These extremely high temperatures are presumably attained by transfer of magnetic energy during violent events. The precise mechanisms of coronal heating remain an area of active research.

At coronal temperatures, the hydrogen atoms which comprise the bulk of the solar atmosphere become stripped of their electrons. The visible corona seen during total eclipses is produced by the scattering of sunlight by electrons in the inner solar atmosphere. Heavier atomic species, such as iron, also become ionized, and their emissions at X-ray wavelengths may be used to trace the corona to great distances outwards from the solar disk. Some streamers in the corona may reach outwards for as much as 100 solar radii, halfway to the Earth.

The ionized gas of the solar atmosphere is an excellent electrical conductor and also has, bound up in it, strong magnetic fields. The subsequent interactions of these magnetic fields with that of the Earth, during their expansion into interplanetary space, can give rise, under certain circumstances, to intensified auroral activity.

4.3.5 Processes in the solar atmosphere

The corona as studied from space is seen to be far from static. Mass ejection events following solar flares or the disappearance of prominences lead to the appearance of denser 'bubbles'—transients—travelling outwards through the corona (Wolfson, 1983). Particularly around the time of sunspot maximum, the inner corona is subject to quite rapid turnover: transients may remove the equivalent of the whole mass of the corona in three months at sunspot maximum (Wentzel, 1989). Following its removal during a transient, coronal gas is quickly replenished from below.

X-ray observations from orbit have been of particular value in confirming the occurrence of several predicted phenomena in the corona. The photosphere appears dark (being too cool to emit at such wavelengths) when the Sun is observed in X-rays. Such observations therefore allow the corona to be observed in its entirety.

The X-ray corona in the direction of the solar disk is seen to contain a number of features. Active regions are revealed as bright in X-rays. Numerous bright points, interpreted as the bases of coronal streamers are seen. Several short-lived bright spots, with durations of the order of eight hours, are also seen against the solar disk: these may give rise to small, but frequent, releases of magnetic flux from the Sun (Forbes, 1994). These events may be important in heating the corona.

Of particular interest are the *coronal holes*, first well visualized from Skylab in the early 1970s. In regions where the magnetic field lines emerging from the Sun are 'open', coronal gas is no longer confined, and can flow outwards and cool relatively freely. These regions therefore have lower gas densities and temperatures than their surroundings, and appear darker in X-rays. These streams of coronal gas sweep outwards into interplanetary space, where they may be encountered by the Earth at regular intervals corresponding to the Sun's rotation. Coronal hole streams can persist for many months, giving rise to recurrent series of quiet aurorae over the period of their existence.

The polar regions of the Sun are thought to be occupied by permanent coronal holes. Seasonal variations in the occurrence of quiet aurorae can be accounted for by the differing presentation of the Earth to the polar coronal hole streams as it orbits: around the equinoxes, the inclination of the Earth's orbit relative to the solar equator brings it more directly into line with the Sun's higher latitudes.

Coronal holes extending to the Sun's equatorial regions tend to appear towards the end of the sunspot cycle, and are the major source of highly magnetized particle streams (and consequent enhancements of terrestrial auroral activity) at this time. The equatorial coronal holes eventually become connected to the polar holes, and gradually retreat polewards as they slowly decay.

4.3.6 Solar flares

While coronal holes are the major source of disturbed geomagnetic conditions in the years leading up to sunspot minimum, by far the most vigorous disturbances result from violent activity in the inner atmosphere of the Sun at those times when sunspots are common. The association between high sunspot numbers and the enhanced likelihood of extensive auroral storms has long been known. Carrington's observation of a white-light solar flare in a sunspot group on September 1 1859, followed a day or so later by major auroral activity, began to cement the connection.

Solar flares occur not *within* the sunspot groups themselves, however, but in the inner corona immediately above active regions. In these locations, intense contorted magnetic field lines are brought into close proximity. As with many solar phenomena, precise mechanisms are still being investigated, but it is widely accepted that solar flares begin from a small, intense 'kernel', within which vast energies are released over short time intervals as a result of magnetic reconnection, and from which shock-waves propagate outwards.

Shock-waves propagating downwards from the kernel to the photosphere give rise to a characteristic two-ribbon structure in major flares. Visualized in hydrogen-alpha light, this often begins life as a series of bright points in an active region, joining to become continuous. Two-ribbon flares apparently mark the footpoints of 'arcades of material either side of a magnetically neutral zone.

Prominences up to 100 000 km away from the site of a large flare may be disturbed by outward-propagating shock waves travelling through the inner solar atmosphere. Particles accelerated to high energies within flare kernels are ejected outwards into the corona.

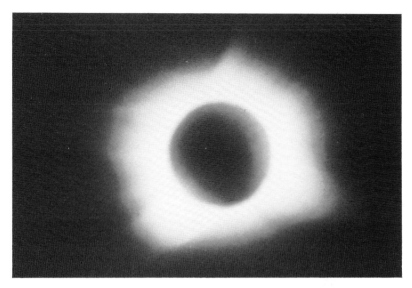

Fig. 4.9. Photograph of the 1983 Java eclipse (with sunspot activity approaching minimum) courtesy of Michael Maunder.

Such ejections are one source of coronal transients. These events may carry a sufficient mass of chromospheric material to add approximately 5% to the solar wind flux (Dryer, 1987).

The onset of a solar flare is marked by an abrupt increase in the X-ray and ultraviolet wavelength emissions from the region in which it is to occur. Characteristically, the rise of a flare to its peak of intensity is more rapid than the subsequent decline. Flares may have durations as short as 10 minutes, or may persist for several hours.

Energetic electrons ejected during solar flares produce bursts of radio noise as they collide with coronal material. Type II radio bursts are detected between 10 and 30 minutes following the onset of a flare, and result from shock-waves travelling outwards at velocities of 800–2000 kms^{-1} through the corona. Electrons accelerated to higher velocities—perhaps as much as half the speed of light—give rise to type III radio bursts. Most energetic are type IV radio bursts, which sometimes follow type II, and are caused by pockets of magnetized plasma travelling out through the corona. These appear to be associated with coronal transients.

Atomic particles accelerated to high energies within solar flares have a number of important effects on the terrestrial environment. Clouds of energetic electrons ejected into the solar wind (section 4.3.8) are involved in producing storm aurorae. Ultraviolet and X-ray emissions during periods of high solar flare activity are also important in heating the upper atmosphere, and increasing its density at higher altitudes, with a concomitant increase in the level of atmospheric drag experienced by low-orbiting satellites (section 8.1).

Flares vary considerably in magnitude (Table 4.1). The most intense solar flares are associated with the most complex active regions, which are often also those containing the most extensive and complex sunspot groups. Such groups are commonest during the early rise of the sunspot cycle towards the maximum of its 11-year cycle, and it is therefore little

Table 4.1. Intensities of solar flares in hydrogen-alpha

Type	Area of solar hemisphere affected (millionths)
Sub-flare(s)	<100
1	100–250
2	250–600
3	600–1200
4	>1200

Flares of each type may be further classified:

F	Faint
N	Normal
B	Bright

Therefore, the most intense flare events will be those of category 4B.

surprise that this is also the time at which the most energetic and extensive flare-induced aurorae are usually seen. Several such events marked the early rise of sunspot cycle 22 during the latter parts of 1988 and the beginning of 1989.

Around sunspot maximum, there may be as many as 25 flares per day on the observable hemisphere of the Sun.

Solar flares are only very rarely visible in white light; most are visualized by routine monitoring of the Sun in hydrogen-alpha wavelengths from ground observatories, or in ultraviolet and X-ray wavelengths from orbiting satellites. As yet, it remains impossible to predict such events more than an hour or so in advance. The onset of a flare is often preceded by disturbances in the magnetic pattern above an active region, but by the time such disturbances are noticeable, the flare itself is often already in progress.

4.3.7 New insights on coronal mass ejections and solar flares

During the early 1990s, an attractive alternative to the 'flare paradigm' presented in section 4.3.6 has been developed, notably by Jack Gosling of Los Alamos National Laboratory and Art Hundhausen of the High Altitude Observatory at Boulder (Crooker, 1994). Gosling and Hundhausen cite coronal mass ejections (CMEs; equivalent to transients) rather than flares as the main class of solar activity responsible for causing major geomagnetic disturbances and low-latitude aurorae.

Modelling of CMEs by Hundhausen suggests that these develop from helmet structures in the corona which become detached from their chromospheric footpoints. Subsequent reorganization of the underlying magnetic field may result (20–30 minutes later) in a flare event: in a reversal of previous dogma, flares may be a consequence of magnetic field reconnection following the CME, rather than vice versa!

Measurements from the ISEE-3 satellite, stationed sunwards of the Earth at the L1 stable orbital point from 1978–82, showed counterstreaming electron currents, characteristic of CMEs, ahead of all but one of the 14 most vigorous magnetic storms in this interval. In

Cycle 22, 36 out of the 37 biggest geomagnetic storms were associated with CMEs, rather than solar flares. Observations at X-ray wavelengths from the Yohkoh satellite in 1992 lend support to the view that CMEs can occur without solar flares as the trigger (Culhane, 1993).

Travelling at velocities which can occasionally exceed 1000 km s^{-1}, well in excess of the undisturbed solar wind (section 4.3.8), these ejections pile into material ahead of them, creating an interplanetary shock wave. CMEs often show a dark trailing edge as they sweep up the slower-moving solar wind in front (Phillips, 1992). The arrival of the shock wave can severely distort Earth's magnetosphere, resulting in geomagnetic storms and their associated lower latitude auroral activity. Further, the 'draping' of field lines ahead of the shock wave may help to satisfy the requirement for a southerly configuration in the interplanetary magnetic field (section 5.4.2.2) to trigger a major geomagnetic disturbance (Kurth, 1992).

Perhaps the future stationing of spacecraft capable of detecting the approach of CMEs by virtue of their counterstreaming currents will afford the possibility of early alerts of disturbed geomagnetic conditions (Appenzeller, 1992): even an hour's warning, offered by a spacecraft at the L1 point 1 500 000 km 'upwind' of Earth might be sufficient to allow satellite operators to shut down or otherwise protect sensitive equipment.

4.3.8 The solar wind

Coronal plasma is largely trapped under magnetic field loops lying above active regions. It has, indeed, been suggested that in the absence of such constraining fields, coronal material would never be sufficiently concentrated to become visible. Solar magnetic field loops reaching to great distances above the photosphere may eventually become so stretched that they become open, however, allowing coronal plasma to flow freely into interplanetary space. Essentially, the kinetic energy of the plasma exceeds the restraining energy of the local magnetic field in such regions. The corona appears to contain several such open regions at any one time, from which long streamers extend past the closed regions delineated by helmet structures (section 4.3.4). Emerging coronal streamers fan out as they extend further from the Sun.

Open regions in the Sun's magnetic field are the source of the solar wind, a continual outflow of plasma from the Sun which permeates the entire solar system. Confirmation of the existence of the predicted solar wind, postulated by E. N. Parker and others into the late 1950s, was one of the early landmark successes of planetary exploration by spacecraft. Equipment aboard Mariner 2, launched towards Venus in 1962, detected the solar wind flowing outwards at 400 km s^{-1}.

The solar wind is not always a steady, quiescent flow. The violent ejections of material associated with large solar flares or the disappearances of prominences lead to increased densities and speeds. Solar flares may introduce turbulent pockets of high solar wind velocities reaching 1000–2000 km s^{-1}. Particle streams from coronal holes produce local solar wind velocities of 800 km s^{-1}.

Completely ionized by the high temperatures in the corona, the material expanding outwards into interplanetary space as the solar wind is a *plasma*, consisting principally of protons (hydrogen nuclei) and electrons, with a small population of other atomic nuclear species. The plasma has, frozen into it, a magnetic field, whose strength and orientation are determined by features at the Sun's surface. The strength and direction of this *Interplanetary Magnetic Field* (IMF) carried along with the solar wind is important in determining the nature and extent of its interactions with the Earth's magnetosphere. In particular, when the IMF has a strong southerly component relative to the ecliptic plane, conditions are favourable for the occurrence of more active and extensive aurorae (section 5.4.2.1).

Emerging plasma initially co-rotates with the Sun below. The solar wind velocity increases with increasing distance from the Sun, as the thermal energy of the plasma overcomes magnetic and gravitational restraints. From 5 to 6 solar radii outwards, the solar wind becomes supersonic. Close to the Sun, but above closed coronal loops, 'open' field lines emerge radially from the Sun. As these field lines are dragged outwards by the solar wind, the rotation of the Sun which carries their points of emergence, begins to wrap them into a spiral. The Interplanetary Magnetic Field becomes increasingly wound into a spiral structure with distance from the Sun. Viewed from above the plane of the ecliptic, the solar wind magnetic field makes an angle of about 45° relative to the direct Sun–Earth line at the distance of the Earth's orbit. The IMF passing the Earth corresponds to the magnetic field of features near the central meridian of the observed solar disk 36–48 hours previously. At the distance of the outer solar system and the heliopause (section 4.3.9), these field lines are directed perpendicular to the radial.

Fine structure in the IMF carried by the solar wind has an important influence on the nature and extent of terrestrial auroral activity and, indeed, that observed in the atmospheres of other planets. On a larger scale, the Sun appears to have semi-permanent regions of magnetic polarity in either hemisphere. These large-scale regions are also evident in the solar wind as magnetic sectors in the plane of the solar equator.

There are usually two or four of these sectors, more or less evenly spaced. Between the hemispheres in the solar wind lies a *neutral sheet*, more or less in the plane of the ecliptic. Solar wind velocities are lowest in the neutral sheet, rising sharply with increasing latitude on either side by an average of 10 km s^{-1} degree^{-1} (Wang *et al*. 1990). Measurements obtained using instruments aboard the Ulysses spacecraft in 1994 also indicated a marked increase in solar wind speed at higher heliographic latitudes. The solar wind emerging from the permanent polar coronal holes has a velocity of 750–800 km s^{-1}.

Around sunspot minimum, the neutral sheet in the solar wind is fairly flat. At times of higher activity, however, it becomes folded into a 'pleated' form, the folds appearing steeper relative to the Sun's equatorial plane at sunspot maximum. The folds of the neutral sheet are, of course, the borders between regions of opposed magnetic polarity in the solar wind. During each solar rotation, folds in the pleated sheet sweep across the Earth from time to time, leading to fairly abrupt changes in the local IMF. These *sector boundary crossings* can lead to short-term increases in geomagnetic activity.

The neutral sheet appears to be largely confined to the lower solar latitudes. Pioneer 11, which passed Jupiter in 1974, was gravitationally slung slightly above the ecliptic plane *en route* to its 1979 encounter with Saturn. During its out-of-ecliptic passage, Pioneer 11 experienced long periods of constant solar wind magnetic polarity, implying that the changes associated with sector boundary crossings and the neutral sheet are indeed principally a phenomenon associated with equatorial heliographic latitudes.

4.3.9 The heliosphere

The solar wind pervades the entire solar system, out to at least the orbit of Neptune, as measured by magnetometers aboard the Pioneer and Voyager spacecraft, now well on their way out of the solar system following successful encounters with the outer planets: in a very real sense, all the known planets can be said to be immersed within the outer fringes of the solar atmosphere!

The volume of space within which the Sun's magnetic influence is dominant is referred to as the *heliosphere*. Early models suggested that this should peter out around the orbit of Saturn, 10 astronomical units (AU) from the Sun. Data from the Pioneers and Voyagers show its extent to be at least 65 AU.

Instruments aboard the Voyager probes detected radio waves at frequencies between 1.8 and 3.5 kHz in July 1992, reaching a peak in December 1992 and declining in early 1993. These low-frequency emissions have been interpreted by Donald A. Gurnett and William S. Kurth of the University of Iowa as being consistent with interactions between material from solar mass ejections (section 4..3.7) and the interstellar medium at the

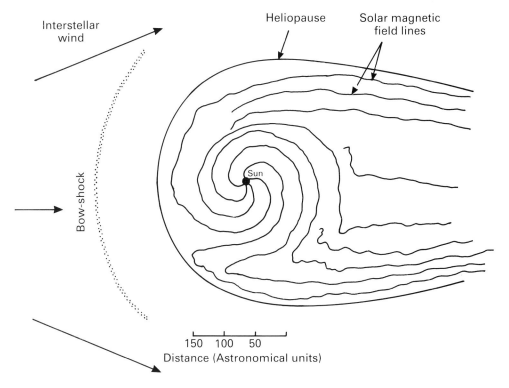

Fig. 4.10. The region of space in which the solar magnetic field, carried outwards with the solar wind, is dominant is termed the *heliosphere*. The heliosphere's precise limits are as yet unknown, but from observations of comets, and data returned by interplanetary probes, its broad structure may be inferred. The solar wind magnetic field initially emerges more or less radially (four sectors in the solar wind are shown), becoming wound into a spiral with distance from the Sun. The postulated intestellar wind draws the heliosphere out into a comet-shaped structure, which may be compared with the Earth's magnetosphere (Fig. 5.1).

boundary of the heliosphere (the *heliopause*). The time delay (1.1 years) between the initial mass ejections and detection of radio noise implies a heliopause distance of 110–160 AU from the Sun (Beatty, 1993).

It seems likely that the heliopause is not at a fixed distance from the Sun, but that it moves inwards and outwards in response to changing solar wind strength at different stages of the sunspot cycle. Long-standing models suggest that, at times of high sunspot activity, the average solar wind velocities should be higher, and the heliosphere consequently more extensive. More recently, Whang and Burlaga have proposed that the long-lived, steady coronal hole streams common at sunspot minimum result in an overall higher dynamic solar wind plasma pressure than that resulting from sporadic interludes of increased pressure (from CMEs, for example) at maximum: the heliosphere may be more extensive at sunspot *minimum*. The issue is one of ongoing debate.

It is possible that the local region of the Galaxy occupied by the solar system is permeated by a 'galactic wind', analogous to the solar wind, but on a larger scale. The heliosphere may be drawn into a comet-shaped structure, with a bow shock in the upwind direction. The solar system as a whole has a proper motion of 20 km s^{-1} towards the constellation of Hercules, and it is in this direction that the heliospheric bow shock might be expected to lie. The Voyagers and Pioneer 11 are leaving the solar system towards the bow shock, whilst Pioneer 10 is heading down the tail of the heliosphere.

High-energy galactic cosmic rays are retarded by the magnetic field of the heliosphere. Beyond the heliopause, it is therefore anticipated that a higher cosmic ray flux should be experienced by instruments aboard the Pioneers and Voyagers. The flux of anomalous low-energy cosmic rays (comprising neutral He, N, O, Ne and H nuclei from interstellar space accelerated at the heliospheric termination shock) is also expected to increase with distance from the Sun (Jokipii and McDonald, 1995). J. A. Van Allen has already reported an increase in observed galactic cosmic ray flux of 1.5–2.0% for every Astronomical Unit travelled outwards from the inner solar system. Whilst, unfortunately, further operation of Pioneer 11 became non-viable late in 1995, data from the three other spacecraft in interstellar space, should they remain operative sufficiently long, are anticipated with interest (Robinett, 1990; Venkatesan and Krimigis, 1990).

REFERENCES

Akasofu, S.-I. (1994) The shape of the solar corona. *Sky and Telescope* **88** (5) 24–27.
Appenzeller, T. (1992) Hope for magnetic storm warnings. *Science* **255** 922–924.
Baliunas, S., and Saar, S. (1992) Unfolding mysteries of stellar cycles. *Astronomy* **20** (5) 42–47.
Baxter, W. M. (1972) *The Sun and the amateur astronomer*. David & Charles.
Beatty, J. K. (1993) Hails from the heliopause. *Sky and Telescope* **86** (4) 30–31.
Bracewell, R. N. (1988) Varves and solar physics. *Q. Jl. R. Astr. Soc.* **29** 119–128.
Crooker, N. (1994) Replacing the solar flare myth. *Nature* **367** 595–596.
Culhane, L. (1993) Yohkoh basks in sunlight. *Nature* **362** 496–497.
Dryer, M. (1987) Solar wind and heliosphere. In: Akasofu, S.-I., and Kamide, Y. (Eds), *The solar wind and the Earth*. D. Reidel.
Forbes, T. G. (1994) 'Fireflies' on the Sun. *Nature* **369** 278–279.

Giovanelli, R. (1984) *Secrets of the Sun.* Cambridge University Press.

Horgan, J. (1991) In the beginning ... . *Scient. Am.* **262** (2) 100–109.

Jokipii, J. R., and McDonald, F. B. (1995) Quest for the limits of the heliosphere. *Scient. Am.* **272** (4) 62–67.

Kurth, W. S. (1992) Tweaking the magnetosphere. *Nature* **356** 18–19.

Leibacher, J. W., Noyes, R. W., Toomre, J., and Ulrich, R. K. (1990) Helioseismology. In *Exploring Space, Scient. Am.* Special issue.

Lockwood, G. W., Skiff, B. A., Baliunas, S. L. and Radick, R. R. (1992) Long-term solar brightness changes estimated from a survey of Sun-like stars. *Nature* **360** 653–655.

Murdin, P. (1990) *End in Fire.* Cambridge University Press.

New, R. (1990) Watching the wobbling Sun. *New Scientist* **125** 54–56.

Nicolson, I. (1982) *The Sun.* Mitchell Beazley.

Noyes, R.W. (1990) The Sun. In: Beatty, J. K., and Chaikin, A. (Eds), *The new Solar System* (3rd Edn). Cambridge University Press.

Phillips, K. J. H. (1992) *Guide to the Sun.* Cambridge University Press.

Robinett, K. H. (1990) The Voyager interstellar mission. *Astronomy Now* **4** (3) 25–28.

Strach, E. H. (1989) *The Astronomer* **25** (300) 244–245.

Venkatesan, D., and Krimigis, S.M. (1990) Into the night between the stars. *Astronomy* **18** (2) 42–-47.

Wang, Y.-M., Sheeley, N. R., and Nash, A. G. (1990) Latitudinal distribution of solar-wind speed from magnetic observations of the Sun. *Nature* **347** 439–444.

Wentzel, D.T. (1989) *The restless Sun.* Smithsonian University Press.

Wolfson, R. (1983) The active solar corona. *Scient. Am.* **248** (2) 86–95.

5

The Earth's magnetosphere

5.1 ORIGIN OF TERRESTRIAL MAGNETISM

The structure of the Earth's deep interior may, of course, only be postulated from indirect observations of phenomena such as the propagation of seismic waves following earthquakes. From such observations, it is has been established that the solid body of the Earth has a many-layered structure (Anderson, 1990; Powell, 1991). The outermost, and thinnest, layer comprises the continental crust. This floats above a deeper mantle, comprised of two principal layers separated by a transition zone in which basaltic magmas originate.

The deepest of all regions, the core, is of about 7000 km diameter (approximately 45% of the overall diameter of the Earth), and is believed to be composed of nickel–iron. The core is under immense pressures (around 3850 kbar), and is at a high temperature, ranging from approximately 3000 K in its outer parts to 6000 K in the centre. The innermost third of the core is solid, and is overlain by a fluid region. Between the two regions of the core, there is probably a relatively thin transition layer.

Electrical currents arising from rapid convective motions in the fluid outer core are the source of terrestrial magnetism. These motions may be driven by small-scale variations in chemical composition, or by the radioactive decay of heavy elements. Energy could also be provided by residual 'primordial heat' generated during Earth's condensation from the protosolar nebula (Jeanloz, 1983). Evidence of terrestrial magnetism is found in the oldest rocks (formed 3.5 billion years ago), implying that the Earth underwent differentiation to its broad current structure fairly rapidly after formation.

The magnetic field generated in the Earth's core has an equatorial strength of 0.3 gauss, having about twice this strength in the polar regions. Studies of palaeomagnetism, the magnetic field imprint frozen into solidifying rocks at past epochs, indicate that the Earth's magnetic field undergoes relatively regular reversals in the long term. The precise locations of the magnetic poles also vary gradually in the short term: movement of the north geomagnetic pole has been invoked as one explanation for the higher frequency of auroral sightings at mid-latitudes in Europe around the twelfth century relative to the present day (section 2.2). Over timescales of centuries, the magnetic field as a whole has drifted westwards (Bloxham and Gubbins, 1989). The 1985 position of the north geomagnetic pole was 79°N 71°W, in the Nares Strait between Ellesmere Island and Greenland: observers at relatively low geographical latitudes in North America enjoy a higher frequency of auroral occurrence than their European colleagues at identical latitudes, by virtue of being closer to the geomagnetic pole.

5.2 STRUCTURE OF THE MAGNETOSPHERE

The magnetic field generated by currents in the Earth's core is extensive. In isolation, it would have a simple dipole configuration, similar to the familiar, symmetrical pattern

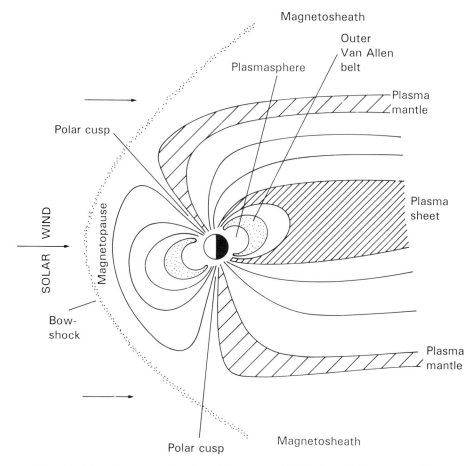

Fig. 5.1. Schematic cross-section through the noon–midnight plane of the magnetosphere, illustrating how magnetic field lines ahead of Earth in the solar wind are compressed, while those in the anti-solar direction are drawn out downwind towards the magnetotail. Most of the solar wind plasma is deflected around the bow-shock, usually 64 000 km upwind of the Earth. Some plasma, however, penetrates via the polar cusps. Plasma is carried downwind along the plasma mantle, and into the neutral plasma sheet lying in the equatorial plane of the magnetotail. Reconnection between solar wind and terrestrial magnetic field lines at the leading edge of the magnetopause—the boundary of Earth's magnetic influence, which lies below the bow-shock—may result in ejection of plasma from the sheet. Plasma is ejected down the magnetotail back into the solar wind, and also into Earth's high atmosphere, there producing the aurora.

produced by field-lines around a schoolchild's bar magnet when iron filings are sprinkled onto a piece of paper laid over it. Interactions between the Earth's magnetic field and the solar wind result in distortion: while 90% of terrestrial magnetism may be represented by a simple, regular dipole, the outermost 10% of the field is pushed earthwards on the day-side, and dragged downwind into a long 'tail' on the night-side. The comet-shaped 'cavity' in the solar wind within which terrestrial magnetism is dominant is referred to as the *magnetosphere*.

Whilst the structure of the magnetic field at ground level may be directly measured quite readily, an understanding of its precise behaviour when extended into near-Earth space had to await the advent of satellite and space probe exploration starting in the late 1950s (section 3.9).

The Earth's magnetic field is a barrier to the solar wind, which streams outwards radially from the Sun at an average velocity of 400 km s⁻¹. About 10 Earth-radii (64 000 km) ahead of the Earth, a shock-wave is produced in the solar wind as this encounters the leading edge of the terrestrial magnetic field. This *bow-shock* has been likened to the wave ahead of the bows of a ship ploughing through water, though the similarity is in some ways only superficial. The bow-shock upstream of the Earth in the solar wind is described as *collisionless*, resulting from field, rather than particle, interactions. These interactions result in heating of the solar wind plasma.

The bulk of the solar wind plasma flow is deflected around the Earth and on into inter-planetary space around the bow-shock without any further interaction with the terrestrial magnetic field. The precise distance of the bow-shock changes in response to conditions in the solar wind: it moves earthwards when the solar wind pressure (velocity) is higher, and sunwards at times of low solar wind intensity.

The boundary of Earth's magnetic influence is delineated by the *magnetopause*, lying under the bow-shock. Between the bow-shock and magnetopause lies a plasma-rich region, the *magnetosheath*.

Terrestrial magnetic field lines in the direction facing towards the Sun are compressed, while those in the anti-solar direction are drawn out towards the distant *magnetotail*, which lies some millions of kilometres downstream in the solar wind, well beyond the Moon's orbital distance. Each month, for a number of days around full, the Moon passes through Earth's magnetotail, which has a cross-section of 40–60 Earth-radii (255 000–382 000 km). It has been suggested that this passage may occasionally have some observable effects (section 8.6).

Viewed in cross-section perpendicular to the Sun-Earth line, the magnetotail comprises two *lobes* of opposed magnetic polarity, derived from the two hemispheres of the terrestrial magnetic field. Field lines emerging from the northern hemisphere are directed sunward, those in the southern lobe anti-sunward. Electric currents in the magnetosheath circulate in opposite senses around these lobes, meeting in the *neutral sheet* which lies in the equatorial plane of the magnetotail. The current in the neutral sheet flows eastward across the magnetotail from the dawn to the dusk side.

While the vast bulk of solar wind material passes the Earth without interaction, some plasma is able to penetrate into the magnetosphere at higher latitudes via the polar cusps. An important route of entry of solar wind plasma is reconnection between Interplanetary

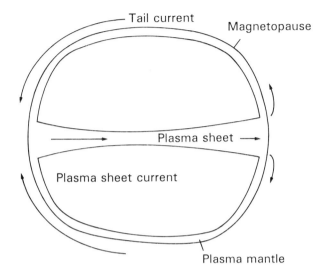

Fig. 5.2. A cross-section through the magnetotail in the dawn–dusk plane, showing the plasma sheet, plasma mantle, and associated currents resulting from plasma movements.

Magnetic Field lines and the terrestrial magnetic field at the 'nose' of the magnetopause (section 5.3). From here, it travels along the outside of the magnetosheath, then enters the *plasma sheet*, a region of hot plasma surrounding the neutral sheet. A much smaller amount of solar wind plasma also diffuses across the magnetosheath at lower latitudes.

Sunlight ionizes the rarefied upper atmosphere, producing the layers of the *ionosphere* (section 6.1.2). Particles from the ionosphere can 'evaporate' upwards along magnetic field lines into the magnetosphere. Almost equal proportions of material in the neutral sheet may be derived from solar wind and terrestrial ionospheric sources. Plasma resident in the neutral sheet can, under certain conditions, be injected into the Earth's high atmosphere, giving rise to auroral activity.

Particles of both solar wind and ionospheric origin can become trapped in certain regions of the magnetosphere close to the Earth, the *Van Allen belts* (section 5.2.1). Lower energy plasma evaporated from the ionosphere populates the *plasmasphere*, which lies under the outer Van Allen belt to a distance of about 4 Earth-radii. The plasma in the plasmasphere is relatively cool by magnetospheric standards, at a temperature of around 2000 K. Material in the plasmasphere co-rotates with the Earth below, unlike the more rapidly circulating particles in the Van Allen belts.

5.2.1 The Van Allen belts

The motions of charged particles in a magnetic field are controlled by a number of influences. Important among these is the *Lorentz force*, which deflects charged particles entering a magnetic field at right angles to the field, and also at right angles to their previous direction of motion. The consequence of this 'right-hand rule' is that particles spiral along magnetic field lines. Being of opposed charge, protons and electrons spiral in opposite senses around a field line.

In certain regions of the magnetosphere relatively close to the Earth, charged particles may become trapped, spiralling back and forth along closed magnetic field lines for long periods. The spirals described by trapped particles are quite open at larger distances from the Earth, where the magnetic field intensity is lower. In the stronger magnetic field closer to the Earth, however, the spirals tighten until a point is reached where the trajectory is exactly perpendicular to the field line. Particles are deflected back along the field line in the opposite direction from this '*mirror point*'.

The trapping regions in the Earth's magnetosphere, predicted by the theoretical studies of Stormer in the early twentieth century, and discovered using radiation detectors aboard Explorer 1 in 1958, are now commonly known as the Van Allen belts.

There are two Van Allen belts. The inner, containing protons and electrons of both solar wind and ionospheric origin, lies around an average distance of 1.5 Earth-radii above the equator. The outer Van Allen belt contains mainly electrons from the solar wind, and has an equatorial distance of 4.5 Earth-radii. 'Horns' of the outer belt dip sharply in towards the polar caps in either hemisphere. As a result of the offset between the Earth's geographic and magnetic axes (section 5.1), the inner belt reaches a minimum altitude of about 250 km above the Atlantic Ocean off the Brazilian coast. This *South Atlantic Anomaly* occupies a region through which low-orbiting satellites frequently pass. Energetic particles in the South Atlantic Anomaly can be a source of problems (section 1.2).

Van Allen belt electrons typically have energies of a few megaelectronvolts, while trapped protons may range from 10 MeV to 700 MeV: fortunately for those using satellites for communications, astronomical observations, or other purposes, protons in the higher range are rather rare.

Particles trapped in the Van Allen belts were once thought to play a primary role in governing auroral activity. It was believed that, when the right conditions prevailed, particles from the Van Allen belts could escape ('precipitate') into the high atmosphere, there giving rise to auroral activity (Jastrow, 1959). This model, however, lacks a satisfactory mechanism for accelerating auroral electrons to the observed energies, and has largely fallen into disfavour. It has also been stressed by some workers that the total particle population in the Van Allen belts would be insufficient to sustain major auroral activity for much longer than an hour or so. The most widely accepted modern theories suggest that the main magnetospheric reservoir of the particles which are subsequently accelerated and injected into the atmosphere to cause the aurora is in the plasma sheet.

In addition to their motion back and forth between the hemispheres, particles trapped in the Van Allen belts undergo a lateral drift, travelling around the Earth in longitude over timescales of minutes to hours: experimental releases and high-altitude nuclear test explosions have demonstrated that this drift rapidly distributes particles around the belts. Being of opposite charge, electrons and protons drift in opposite directions around the Earth: electrons drift eastwards, protons westwards.

The main influence of the Van Allen belt particle populations during auroral activity, according to more recent theories, is in decreasing the global magnetic field intensity, as they circulate faster in response to disturbed conditions. Ring currents in the Van Allen belts intensify markedly at the onset of geomagnetic storms, and contribute to disturbances of the global ground-level magnetic field at such times.

Particles in the Van Allen belts are trapped for long periods, but there is some degree of turnover, and they may be lost from the belts by several processes. Radio waves, from natural sources such as lightning discharges (producing 'whistlers') or man-made sources, can transmit along closed terrestrial magnetic field lines looping out through near-Earth space. Resonances between these waves and trapped particles can reduce the particles' kinetic energies, and cause them to precipitate out of the trapping regions. More importantly, resonances may increase particle velocities along the field lines relative to their perpendicular velocities, resulting in the mirror point being moved closer to the Earth; if the mirror point is moved sufficiently close, particles will hit the atmosphere.

Neutral hydrogen (unaffected by magnetic field effects) escapes from the upper atmosphere and permeates near-Earth space to distances of a few Earth-radii to produce the *geocorona*. Collision between trapped particles and geocoronal hydrogen is another mechanism by which material is lost from the Van Allen belts.

The particle concentration is also subject to change, largely in response to geomagnetic activity. Particle concentrations in the inner belt change only very slowly, by a factor of about three over the course of 12 months or so. The outer Van Allen belt particle population may change by a factor of ten in less than a day.

5.2.2 A new radiation belt

Launched into a 550 km altitude orbit in June 1992, the Solar Anomalous and Magnetospheric Particle Explorer (SAMPEX) carried four particle detectors, designed to sample energetic ion and electron populations in near-Earth space. SAMPEX detected a third belt of energetic trapped particles, lying between the previously known Van Allen belts (Allen, 1993).

The belt detected by SAMPEX is strongest over the Atlantic between Africa and South America. The trapped particles are believed to be extra-solar in origin (unlike the solar wind-derived material forming the Van Allen belts). Measurements obtained by CRRES during auroral storms in March 1991 indicated the presence in this region (6000–7000 km above Earth's surface) of singly ionized oxygen, derived from anomalous cosmic rays (section 4.3.9). The population in this third radiation belt is therefore expected to be modulated by the solar cycle, being denser at sunspot minimum as the solar wind weakens, allowing such cosmic rays to penetrate further into the inner solar system.

5.3 PLASMA MOVEMENTS AND CURRENTS IN THE MAGNETOSPHERE; THE AURORAL OVALS

Geographically, auroral activity is present more or less permanently at high latitudes, around the *auroral ovals*. These are rings of aurora, asymmetrically displaced around either geomagnetic pole. Under quiet geomagnetic conditions, the auroral ovals remain relatively fixed in space, above the rotating Earth. Each auroral oval extends furthest towards the equator on the night-side, such that an observer at a given mid-latitude geographical location is carried closest to the oval around the time of magnetic midnight. On the day-side, the auroral ovals appear pushed back towards the geomagnetic poles, reflecting the overall distortion of the magnetosphere resulting from its interaction with the solar wind.

Solar wind plasma entering the magnetosphere carries with it a frozen-in magnetic field derived from the IMF. An important process in driving the auroral mechanism is joining—*reconnection*—between the IMF and terrestrial field lines at the leading edge of the magnetopause upwind of the Earth. This merging process occurs with highest efficiency when the IMF is directed southwards relative to the ecliptic plane—that is, anti-parallel to the terrestrial magnetic field. A crude analogy is the joining together of two bar magnets by their opposed (N and S) poles. The reconnected field lines join the magnetosphere to the solar wind, resulting in transfer of energy across the magnetopause boundary.

Such joined field lines, and their entrained plasma, are dragged down the magnetotail by the solar wind. Plasma flows along the outside of the magnetotail lobes in a *plasma mantle* on the boundary of the magnetosheath. As it travels, the plasma is subject to the influence of electric currents in the magnetotail, experiencing *E-cross-B drift* perpendicular to the current and to the magnetic field. Both electrons and protons are deflected in the same sense, so the plasma flows together as an ensemble. The plasma undergoes a cross-field drift and is directed towards the equatorial neutral sheet, where the opposed magnetic polarities of either hemisphere meet and cancel.

Frozen-in magnetic fields will move with a plasma population provided there is no sudden change in local magnetic field direction or drop in its intensity. Where abrupt changes or weakening do occur, magnetic field lines are able to diffuse through the plasma and the frozen-in condition breaks down. Where such diffusion occurs, field lines of opposed magnetic polarity can come together and reconnect. Such a diffusion region exists in the magnetosphere, lying at a distance of some 100 Earth-radii (637 000 km) downwind in the magnetotail under undisturbed conditions: it is estimated that the effective boundary of the terrestrial magnetic field would lie at this distance were it to exist in isolation, rather than being embedded within the solar wind. Reconnection in this region produces a *neutral line*.

'Interplanetary' field lines, reconnected at the neutral line, sweep on down the magnetotail to rejoin the solar wind. Plasma jets are ejected both upstream and downstream from the neutral line. The jet directed upstream, back towards the Earth, helps to maintain the population in the plasma sheet.

The same cross-field currents which influence the movement of plasma in the magnetotail are also projected onto the auroral ovals, such that there is a general current (electron flow) descending onto the evening sector, across the oval, and back out into the magnetosphere on the dawn side (Akasofu, 1989). A lesser, secondary, current flows in the opposite direction. These currents flow along magnetic field lines, and the auroral ovals could, effectively, be considered as extensions of the magnetosphere reaching down into the ionosphere in the Earth's upper atmosphere. Under undisturbed conditions, the auroral ovals are quite narrow, containing thin sheets of discrete aurora (perhaps of the order of 1 km wide), surrounded by a much broader region (500 km or so) of diffuse aurora. During substorms and geomagnetic storms, the ovals become broader, and more complex in structure.

Quiet aurora at high latitudes is produced by a steady stream of particles precipitating out of the earthward extension of the plasma sheet around the polar cap. This quiet aurora may, from time to time, be enhanced by an increased particle flux into the atmosphere.

Such increases occur at the time of *magnetic substorms*, fluctuations in the geomagnetic field resulting from frequent, small changes in the solar wind IMF in the near-Earth environment.

Much more dramatic are the *geomagnetic storms* which result from major variations in IMF intensity and direction. During these events, which usually follow violent activity in the inner solar atmosphere (sections 4.3.5, 4.3.6 and 4.3.7), the auroral ovals become greatly disturbed, broadening and expanding equatorwards, particularly on the night-side. Such events bring the aurora to the skies of middle latitudes. While substorms are of fairly short duration (a few hours), it may take several days for the magnetosphere to settle down following a major geomagnetic storm.

Magnetic substorms and geomagnetic storms represent periods during which the power of the magnetosphere–ionosphere circuit increases, resulting in brighter and more extensive auroral activity. Several commentators have used the analogy of a television set to describe this situation, with the upper atmosphere as the 'screen' illuminated by electrons fired from the magnetotail 'gun'. The shifting pattern of auroral activity is the signature of large-scale interactions between the magnetosphere and the solar wind.

Electric currents flow between the outer and inner edges of the auroral ovals, parallel to the Earth's surface. As a result of the sharp dip of the magnetic field at the high latitudes of the quiet-condition auroral ovals, these currents are directed *perpendicular* to the magnetic field, giving rise to an E-cross-B drift, with protons and electrons drifting from the day-side to the night-side.

At higher altitudes (300 km or so), protons and electrons drift at the same rate. Lower in the ionosphere (around 100 km altitude), the atmospheric density is higher, leading to a selective loss (via particle collisions) of protons. The result of this process is the establishment of electric currents flowing eastwards along the evening sector of the auroral oval, and westwards in the morning sector. These *auroral electrojets* in the ionosphere become enhanced at times of high geomagnetic activity. Where the eastward and westward electrojets meet, around the midnight point on the auroral oval, lies a region of turbulence, the *Harang discontinuity*.

The visible aurora results from collisions between energetic electrons and atmospheric particles. Electrons from the neutral sheet plasma carry insufficient energy to penetrate the atmosphere. As they travel along magnetic field lines into near-Earth space, these are subjected to the same mirroring experienced by electrons trapped in the Van Allen belts. Following the 1977 multi-spacecraft ISEE mission, however, a mechanism by which plasma sheet electrons could become accelerated, the *auroral potential structure*, became apparent.

The auroral potential structure is produced at times when the power of the magnetosphere–ionosphere circuit increases, as during magnetic substorms or geomagnetic storms. Thin sheets of positive and negative charge, aligned to magnetic field lines and lying close together, develop. Acceleration of electrons along these thin sheets offers an explanation for the aurora's appearance in narrow 'draperies' in the high atmosphere. Over distances of 10 000–20 000 km (1.5–3.0 Earth-radii), a potential drop of the order of kilovolts arises, down which electrons can be accelerated, gaining enough energy (several kiloelectronvolts) to penetrate the atmosphere, and there produce the visible aurora.

5.4 INTERACTIONS WITH THE SOLAR WIND

5.4.1 Comets

The most obvious interactions between the solar wind and small solar system bodies are probably those observed in comets. Comet–solar wind interactions show some parallels with those between planetary magnetospheres and the solar wind.

The Giotto spacecraft encounter with Comet Halley on March 13 1986 broadly confirmed the 'dirty snowball' cometary model proposed in the 1950s by Fred Whipple. The nucleus of Halley is an irregular body comprising dusty material bound up in an icy matrix, with a thin outer crust. At great distances from the Sun, typically beyond the orbit of Saturn, such bodies are inert, and essentially indistinguishable from asteroidal objects. Within about 3 AU from the Sun, sublimation of cometary gases occurs in response to heating by solar radiation (Brandt, 1990a).

Initially confined to a small coma immediately surrounding the nucleus, material ejected from the comet soon becomes drawn out into tails by the action of the solar wind and radiation pressure. Dusty material forms a curved tail, as the small debris falls away to pursue an independent orbit around the Sun as a meteor stream. Gas, ionized by the ultraviolet component of solar radiation, forms narrower, more or less straight tails generally pointing directly away from the Sun. Fluorescence at 420 nm wavelength by carbon monoxide (CO^+) ions released into the coma is particularly marked in some comets, and allows the bluish plasma (or ion) tails to be quite easily traced.

The accepted mechanism by which cometary ion tails are produced is essentially that proposed by Hannes Alfven in the late 1950s. Ions produced by the action of solar ultraviolet on the coma become trapped along field lines of the IMF in the passing solar wind. Consequently, the solar wind is decelerated in the comet's neighbourhood. Close to the nucleus, where the cometary ion concentration is highest, the solar wind magnetic field is excluded, and a cavity is formed. Interplanetary magnetic field lines become draped over the cavity's boundary, which is referred to as the *contact surface*. During its encounter with P/Halley, Giotto detected a contact surface 3600–4500 km from the nucleus. Within this region, the ion signature recorded using particle detectors changed from a mixture of heavy cometary ions and solar wind protons, to that of cometary ions alone.

Lobes of opposed magnetic polarity are produced along the comet tails, similar to those seen in planetary magnetotails. Visible ion tails lie along the neutral current sheet between these lobes.

Like planetary magnetospheres, comets have, ahead of them in the solar wind, bow-shocks, while disturbances can extend for considerable distances downwind. Cometary bow shocks are less clearly defined than those produced by planetary magnetospheres. The distance from the nucleus of the bow shock is governed by the comet's rate of gas production. Measurements from the ICE spacecraft at P/Giacobini-Zinner in 1985, and Giotto and other probes at P/Halley in 1986 showed the P/Halley bow shock to lie ten times further (about 1 million kilometres) from the nucleus; P/Halley was estimated to be 30 times as productive of gas as P/Giacobini-Zinner (Coates and Mason, 1992).

Following a period of 'hibernation' after the P/Halley encounter, Giotto was targeted to a further comet, P/Grigg-Skjellerup, which it encountered on 1992 July 10. P/Grigg-Skjellerup is considered to be older, probably smaller, and certainly much less active than

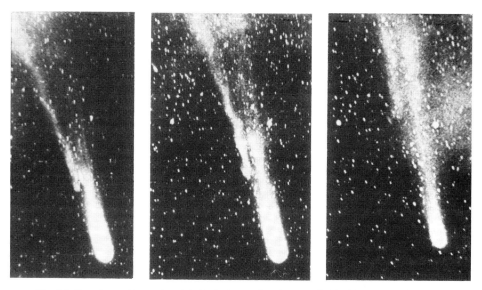

Fig. 5.3. Comets provide good models for the interactions between the Earth's magnetosphere and the solar wind. In particular, interactions between the interplanetary magnetic field and cometary ion tails may produce disconnection events, as seen here in Lowell Observatory photographs of P/Halley at its 1910 return, in which ion tails which had been developing in one direction (determined by the interplanetary magnetic field) become sheared off, to be replaced by a new tail developing in a different direction.

either P/Giacobini-Zinner or P/Halley, and much depleted in the ices involved in gas production. During the P/Grigg-Skjellerup encounter, Giotto ran across a bow shock around 20 000 km from the nucleus. The comet's gas production rate was probably a hundredth that of P/Halley (Bond, 1993).

The interactions between the ion tails of comets and the IMF carried by the solar wind are of particular interest. Changes in the intensity and direction of the IMF give rise to disconnection events, in which ion tails which had been developing in one direction may become sheared off, following which new ion tails reflecting the changed IMF direction in the comet's vicinity begin to grow. Such events may result from the crossing of sector boundaries in the neutral sheet of the solar wind (section 4.3.7) or from encounters with pockets of disturbed solar wind introduced by solar flare events. Ion tail disconnection events involve magnetic field line reconnections, beginning upwind on the sunward side of the nucleus, and are fairly gradual, taking 12 hours or more to complete.

Comets might be regarded as natural probes of regions of the outer solar atmosphere to which Earth-based observers have no other access, as was recognized by Biermann in the early 1950s. Whipple has given the apposite description of comets as solar wind-socks'.

Several disconnection events were observed during the perihelion passage of Comet Halley in 1986, mostly in association with sector-boundary crossings (Brandt, 1990b). Similar correlations were found by Brandt and Niedner during the perihelion passage of Comet Kohoutek in 1973–74. Despite the proximity of the Halley apparition to sunspot minimum, the early months of 1986 proved to be a time marked by a 'gusty' solar wind, which may have served to some extent to increase the comet's nuclear activity.

It has also been proposed that solar activity was responsible for a remarkable brightening of P/Halley observed in February 1991, when the comet lay 14.3 AU from the Sun (far beyond the orbit of Saturn) at a distance where the nucleus would be expected to be dormant. Shock waves associated with solar flares at the end of January 1991 may have arrived in the comet's vicinity some two weeks later, acting as a trigger for unusual activity (Intriligator and Dryer, 1991).

5.4.2 The Earth's magnetosphere and the solar wind

The detailed interactions between the Earth's magnetosphere and the interplanetary medium are complicated. It is to these interactions, however, that we must look to understand the generation of the aurora. Aurorae occur in response to the arrival of energetic, magnetized plasma from the solar wind. This adds to, and may disturb, the existing plasma population in the magnetotail. The associated magnetic interactions may then lead to injection of large numbers of accelerated electrons into the high atmosphere, where they cause excitation of oxygen and nitrogen. In shedding the excess energy of excitation, the atmospheric particles release the light of the visible aurora (section 6.2).

The most direct route of entry for solar wind plasma is via the narrow *cusps* between field lines on the day-side of the magnetosphere at high latitudes. In these regions, the magnetic field lines are closely bundled together, and dip sharply towards the Earth's surface. Particles dribbling in by this route carry relatively little energy and consequently give rise only to diffuse, fairly pallid aurorae (section 6.4.1). The structure of the cusp regions has been clarified by satellite observations and measurements. Surrounding either cusp is a broader region, the *cleft*, which is also characterized by low-energy precipitation, but does possess some higher energy structural features.

A matter of some debate is whether the arrival of solar wind plasma via the cusps is steady and continuous, or erratic. Radar and satellite studies of the ion population in the arriving plasma allows the two possibilities to be distinguished. Ions are separated in time and latitude of arrival by their energies: higher energy ions arrive first, and at lower latitudes. It is therefore expected that, if merging between magnetospheric and interplanetary field lines were continuous, ion energies close to the cusp would show a continuous dispersion in latitude. If, however, reconnection occurs in bursts, then ion dispersion should be discontinuous.

Observations obtained using EISCAT and a DMSP satellite in 1992 show the ion distribution to be consistent with discontinuous arrival, described as a 'pulsating cusp' model (Lockwood *et al.* 1993). Discrete plasma structures, characteristic of intermittent merging between the IMF and terrestrial field lines at the subsolar point on the magnetopause, were observed to move polewards to the cusp.

Solar wind particles entering the magnetosphere are also distributed into the plasma sheet, lying in the equatorial plane of the magnetotail. Reconnection of field lines in the magnetotail at times when the IMF turns southwards relative to the ecliptic plane, results in particles from the neutral sheet becoming accelerated, and travelling earthwards. Arrival of these energetic particles—in effect, the magnetosphere could be viewed as a huge, natural particle accelerator—has several consequences, including the production of more vigorous, often colourful auroral activity, with displays penetrating to lower latitudes.

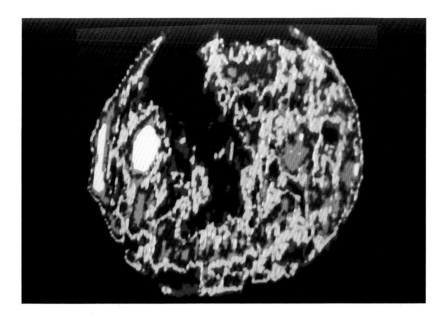

Plate 1 — An extreme ultraviolet image of the Sun, taken from Skylab on 19 August 1973, and showing the 'Boot of Italy' coronal hole. Coronal holes are sources of persistent high-speed streams in the solar wind, and can give rise to recurrent, relatively quiet aurorae. National Space and Aeronautics Administration/Jet Propulsion Laboratory photograph 73–HC–752.

Plate 2 — Progress of an auroral substorm, recorded in a series of images taken from the US Dynamics Explorer 1 satellite near apogee. Images were recorded at 12-minute intervals. Photograph courtesy of University of Iowa, Department of Physics and Astronomy.

ATLAS OF AURORAL FORMS (PLATES 3–14)

Plate 3 — Auroral *glow*, comprising diffuse auroral light with no apparent structure. This may appear visually colourless (particularly if faint) or sometimes slightly bluish-green; as with other photographs in this section, the sensitivity of the film has emphasized the aurora's colour. Photograph from Edinburgh by Dave Gavine, 1989.

Plate 4 — On occasion, the aurora may appear as discrete *patches* (sometimes called *surfaces*), as in this display photographed from Fort Augustus in the Scottish Highlands on 27–28 August 1978 by Dave Gavine. This patch of auroral light brightened and faded over a timescale of several minutes, and appeared to be a section of an incomplete arc (e.g. Plate 5). Such auroral forms undoubtedly account for many spurious 'UFO' sightings.

Plate 5 — As an auroral display increases in activity, the glow may rise higher into the sky, taking on the discrete form of an *arc*. This arc, containing no further internal structure, is *homogeneous*. Note the sharp lower border and dark sky below the arc, compared with the diffuse upper regions. Photograph taken in August 1989 by Tom McEwan, Ayshire, Scotland.

Plate 6 — The same arc as in Plate 5, but now showing increased activity. Short *rays* have developed along its length, producing a rayed arc (mR_1A in the standard abbreviated reporting code described in Chapter 7). Photograph taken in August 1989 by Tom McEwan, Ayrshire, Scotland.

Plate 7 — *Rayed band*, with short rays (mR$_1$B) photographed from Chichester, West Sussex, England by the author during the major auroral storm of 8–9 November 1991. Green oxygen emissions predominate in this picture.

Plate 8 — Another example of a rayed band, again photographed from Chichester by the author, on 24–25 March 1991. Dark sky can be seen below the base of the aurora, and the rays are moderately long (mR$_2$B), reaching into the stars of Cassiopeia and Perseus.

Plate 9 — Rayed structure—long rays (mR_3R)—from a rayed band during the extensive display of 25–26 April 1989. The differing oxygen emissions at higher and lower levels in the atmosphere are reflected in the colour contrast: red emissions occur at great heights at the ray-tops, while green (557.7 nm wavelength) emission predominates nearer the base of the aurora. Photograph by Dave Gavine, Edinburgh.

Plate 10 — Wide-angle lenses can be used effectively to photograph the most extensive displays. Dave Gavine used a 16-mm lens to record the aurora of 28–29 July 1990, which showed complicated multiple band structures, and rays filling much of the northern (poleward) sky from Edinburgh. Hints of the violet-purple (391.4 nm wavelength) N_2^+ emission, resulting from resonance excitation of molecular nitrogen by solar ultraviolet, can be seen.

Plate 11 — A more pronounced example of the violet-purple N_2^+ emission in a rayed band photographed on 29–30 April 1990 by Dave Gavine from Edinburgh.

Plate 12 — The major aurora of 24–25 March 1991 was visible from the south of England, including Chichester (whose Cathedral spire is visible at bottom right). During one of the display's more active phases, the author captured the pronounced red oxygen emission in the long rays to the northeast of the sky. Again, this has been enhanced by the photographic emulsion's high sensitivity.

Plate 13 — Another major auroral display showing the strong red oxygen emission at high altitudes contrasting with green emission lower down, occurred on 8–9 November 1991. The author's photograph from Chichester shows long rays (mR_3R) reaching to the Pole Star.

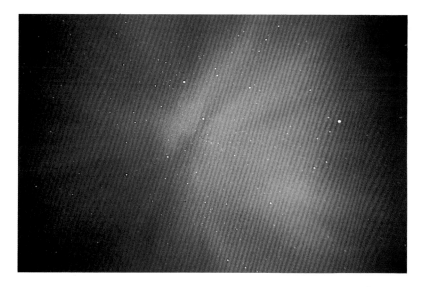

Plate 14 — In extremely active displays, and rather rarely at mid-latitudes, the aurora may fill the whole sky, and assume the form of a *corona* centred on the observer's magnetic zenith. The corona is a perspective effect. This stunning example photographed by Richard Pearce from Elgin, on 21–22 October 1989, appears against the backdrop of the 'Summer Triangle' stars of Cygnus and Lyra.

Plate 15 — Although superficially similar to cirrus, *noctilucent clouds* occur much higher in the atmosphere, at altitudes in excess of 80 km. Consequently, they remain sunlit on summer evenings long after sunset, and appear bright against the twilit background. This fairly typical display was photographed from Cambridge, England, by the author on 25–26 June 1986.

Plate 16 — Rarely, noctilucent cloud displays can be extremely bright and extensive. One such instance was the major occurrence of 23–24 July 1986, seen widely over the British Isles, and photographed here by Dave Gavine from Edinburgh.

5.4.2.1 *Auroral substorms*

For much of the time, the aurora at high latitudes is relatively quiescent, in the form of narrow, faint arcs. Occasionally, activity may suddenly intensify and brighten, during auroral substorms. The occurrence of such events as global phenomena first became recognized by Akasofu and his colleagues, following careful study of simultaneously taken photographs from ground stations, obtained during the IGY. The overall pattern of activity around the oval during substorms has been confirmed by satellite images from polar orbit.

The onset of substorm aurora is quite rapid, and first manifests as a brightening of the oval during what is termed the *growth* phase. Activity spreads both eastwards and westwards on the night-side of the auroral oval during the subsequent *expansive* phase, then begins to migrate polewards, resulting in the production of a wavy structure which moves westwards from the midnight sector into the dusk sector of the oval. This *westward travelling surge* moves at about 1 km s^{-1}. Meanwhile, the morning sector of the oval breaks up

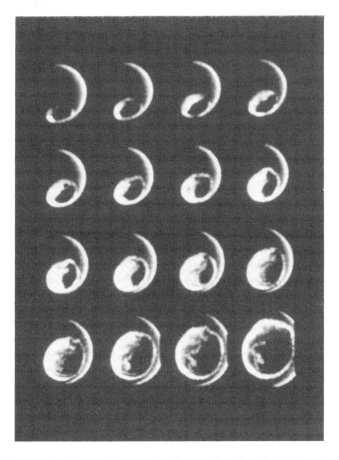

Fig. 5.4. Progress of a substorm in the auroral oval as seen from above by the Dynamics Explorer 1 satellite. Images, taken in the ultraviolet, at intervals of several minutes show the initial brightening of the night-side of the oval, followed by the westward-travelling surge, and expansion of activity to higher latitudes within the oval. University of Iowa photograph.

into patchy rayed structures. The expansive phase may last for some 30 minutes. Once the polewards migration has reached its maximum extent (75–80° geomagnetic latitude), activity begins to subside, during the *recovery* phase (Akasofu and Kamide, 1987).

Substorm aurorae are seen only at high latitudes, and result from relatively smaller geomagnetic disturbances than the major storm aurorae occasionally seen at mid-latitudes. There may be as many as four or five substorms, lasting between 1 and 3 hours, each day.

While the precise details remain to be completely understood, it is now widely accepted that the main driving force behind substorm activity is the condition of the solar wind in the Earth's vicinity. Substorm activity reflects the efficiency of coupling between the IMF and the terrestrial magnetic field. At times of most efficient coupling, the aurora brightens and substorm aurora is seen. Akasofu has derived an equation to describe the efficiency with which the auroral dynamo is driven:

$$P = VB^2 \sin^4 (\theta/2) \, l_0^2 \tag{5.1}$$

where V is the solar wind velocity, B is the strength of the IMF, is the polar angle of the IMF, and l_0 is a distance constant for which 7 Earth-radii seems a reasonable value.

Of these factors, θ appears to be most critical. Particularly if it has a strong southerly component, θ gives rise to a significant increase in the power of the auroral dynamo. Under normal conditions, when the Sun is relatively undisturbed, IMF intensity, B, remains relatively constant (about 5 nT). Solar wind velocity can vary considerably, but θ still plays a more important role in driving the substorm aurorae: V is a more important factor with respect to geomagnetic storms. It is notable that even quite major disturbances in which θ turns northwards are inefficient in producing auroral activity, perhaps offering an explanation as to why not all solar flares are followed by extensive auroral activity.

Magnetospheric substorms are apparently caused by the release of excess energy, some of which is directed earthwards to cause high-latitude aurorae. The power (P in equation (5.1)) generated by the auroral dynamo may be of the order of 10^6 megawatts, and is associated with an electric potential of 100 000 volts (Akasofu, 1982). An important fraction is also ejected down the magnetotail and back into the solar wind in the form of *plasmoids*, magnetic 'cage' structures enveloping significant quantities of magnetospheric plasma, discovered as a result of studies using the ISEE-3 spacecraft in 1983 (Hones, 1986).

Early models for the generation of magnetic substorms invoked spontaneous, explosive reconnection between opposed magnetic field lines, induced by the circulating currents in the lobes of the magnetotail. More recently, Edward W. Hones has proposed a model relating substorm aurora to periods of high southerly directed IMF. The onset of substorm activity is preceded by progressive stretching of magnetic field lines lying more than 7 Earth-radii (l_0 in Akasofu's equation) downstream of the Earth. This stretching is a consequence of more efficient reconnection between IMF field lines and field lines on the sunward leading edge of the magnetopause. In turn, the magnetotail acquires an excess of (solar wind-derived) energy.

The key event seems to be the spontaneous (and as yet unexplained) development of a new neutral line—the *substorm neutral line*—at a distance of about 15 Earth-radii downwind from the Earth. This substorm neutral line interferes with the cross-tail electric current in the neutral sheet, resulting in an abrupt collapse of the stretched magnetic field structure which has developed over the previous hour or so. This collapse is accompanied by the deposition into the atmosphere of large numbers of accelerated electrons from the magnetosphere, which give rise to enhanced auroral activity on collision with atmospheric oxygen and nitrogen.

5.4.2.2 Geomagnetic storms

Auroral substorms at high latitudes arise when the magnetosphere is moderately disturbed, in response to relatively small changes in the solar wind IMF. The more major distur-bances, which follow violent events associated with flares in the inner solar atmosphere, have more dramatic consequences. The geomagnetic storms which sometimes (but do not *always*) follow these events are much less frequent than the high-latitude substorms, but can last for several days. Geomagnetic storms carry auroral activity to lower latitudes, and into the skies of the more populous regions of the world. These are the events witnessed by our forebears as 'battles' or 'dragons' in the sky, and of which the Great Aurorae of 1909, 1938 and 1989 are further, modern examples (sections 2.1, 2.6, and 2.7).

As previously noted, such displays are commonest in the rising phase of a sunspot cycle towards maximum. At these times, the large, actively changing sunspot regions prone to flare production are often present on the solar disk. It should also be noted, however, that such violent events are not unknown in the years close to sunspot minimum.

Geomagnetic storm activity can, again, be understood in terms of Akasofu's equation (5.1). As with substorm activity, the polar angle, θ, of the IMF plays a key role in the generation of geomagnetic storms. Other elements of the equation, however, also come markedly into play.

Coronal mass ejections (sections 4.3.6 and 4.3.7), travelling out from the Sun at veloci-ties of up to 2000 kms^{-1}, give rise to shock-waves in the solar wind. Compression of the magnetosphere by such shock-waves intensifies the terrestrial magnetic field for a period of some tens of minutes, producing an effect detected using magnetometers at ground level as *Sudden Storm Commencement* (SSC). The onset of SSC can be an early indicator of the possibility of aurora visible at lower latitudes, but not all SSC events are necessarily followed by strong visible displays.

One consequence of the compression of the magnetosphere at the time of SSC is an intensification of the ring currents circulating in the Van Allen belts. The compression effect of impact between the magnetosphere and a shock-wave propagating out through the solar wind also acts to intensify magnetic fields in the solar wind. Values of B, the IMF intensity, therefore become elevated. Coupled with the 4–5-fold increase in the value of V produced by the coronal mass ejection, this serves to raise the importance of the function VB^2 in equation (5.1).

In the production of geomagnetic storms, however, the most significant element of Akasofu's equation remains the IMF polar angle, θ. Even very major ejection events will have little or no effect (other than production of SSC) if it is directed northwards, and the efficiency of reconnection between IMF and terrestrial magnetic field lines is low. Even on occasions when the shock is accompanied by a southerly directed IMF, it appears to be necessary for such conditions to prevail for many hours (Akasofu suggests a minimum of six hours) before a full-scale geomagnetic storm will be initiated.

The direction from which the shock-wave in the solar wind approaches the magnetosphere is also important. Flares associated with sunspot groups near the solar limb eject shocks more or less at right angles to the Sun–Earth line. The shock-wave in such instances will glance mostly *across* the leading edge of the magnetosphere, causing little compression, and with little chance of initiation of major auroral activity. If, however, the shock-wave comes more or less head-on towards Earth down the Sun–Earth line, the compression may be very major, and be followed by extensive auroral activity.

This was, indeed, the situation in March 1989. Major flares associated with the progenitor spot-group occurred while the group was close to the limb: some flare-associated ejections were even detected while the spot-group was on the Sun's averted hemisphere *behind* the limb! These had little effect on the Earth. A week or so later, when the group was near the Sun's central meridian as viewed from Earth, another flare occurred, whose associated shock-wave arrived, head-on, within 24–36 hours, producing the best auroral display seen for over 50 years!

Under these conditions, the auroral ovals again brighten as the power of the auroral dynamo increases. A westward-travelling surge sets in, and the ovals expand towards the pole. This expansion is also accompanied, in geomagnetic storms, by an expansion *equatorwards* from the ovals, and the formation of multiple bright arcs on the night-side. The whole structure can become very extensive, bringing the aurora into the skies of mid-latitudes (section 7.2.1). 'Gusty' conditions in the solar wind behind the flare-associated shock-wave may be manifested in the sometimes chaotic moment-to-moment changes in the nature of auroral activity during these storms.

Events associated with solar flares increase velocities and magnetic field intensities in the solar wind. The localized particle density is also increased, from an estimated quiet solar wind value of 5×10^6 m^{-3}, to 1×10^7 m^{-3} (Ratcliffe, 1972).

5.4.2.3 Coronal hole aurorae

The late-cycle persistent streams ejected into the solar wind from coronal holes (section 4.3.5) give rise to geomagnetic activity in some respects intermediate between that of substorm and geomagnetic storm aurorae. As with the latter, a key element appears to be the velocity component. Coronal hole streams can typically reach velocities of 800 km s^{-1}. Compression effects as a coronal hole stream sweeps across the leading edge of the magnetosphere again lead to localized intensification of the IMF in the solar wind, in turn leading to enhanced auroral activity.

The passage of coronal hole streams brings about equatorwards expansion of the auroral ovals, though not to the extreme extent seen during geomagnetic storms. Also, the IMF in coronal hole streams is less prone to momentary changes or turbulence, such that auroral activity is also less violent. Characteristically, coronal hole aurorae penetrate to higher mid-latitudes (and are rarely, if ever, seen from, say, the latitudes of London or the southern United States), and consist of quiet arc or band structures, with little of the rapidly moving rayed activity seen during major geomagnetic storms. The intensity of coronal hole displays is also lower, with fainter, colourless forms predominating.

Coronal holes may persist, relatively undiminished, for many months, with the resulting possibility of recurrent activity each solar rotation, as the stream sweeps across the Earth repeatedly. Such series of quiet aurorae at higher mid-latitudes, recurring every 27 days, are indeed common in the year or so before sunspot minimum, when the coronal holes are most common at lower heliographic latitudes.

As a result of the tilt of the Earth's orbit relative to the solar equator, a seasonal effect in the occurrence of coronal hole aurorae also operates. Earth is more closely aligned with coronal holes at more typical, higher, heliographic latitudes around the equinoxes. Coronal hole-associated aurorae are therefore most often seen around March and September.

5.4.2.4 Sector boundary crossings and chromospheric filament disappearances

At fairly regular intervals, the rotation of the Sun carries broad, large-scale sectors of opposed magnetic polarity across the near-Earth environment (section 4.3.8). A consequence of this can be a rapid change in the local IMF, leading to periods of somewhat enhanced geomagnetic activity, similar to those which arise from the passage of coronal hole streams. The effects of these sector boundary crossings are most noticeable at times of otherwise low activity, being masked at sunspot maximum by the turbulent conditions then prevailing in the solar wind.

The disappearance of chromospheric filaments (equivalent to solar prominences, but seen—in the wavelength of hydrogen-alpha light—as dark against the Sun's disk) can also result in an enhancement (often relatively small) of the IMF passing the Earth, and be followed by increased geomagnetic and auroral activity. Again, these effects are most obvious in the declining years of the sunspot cycle. Disappearing filaments offer an explanation for some of the reasonable late-cycle auroral displays which can extend to lower latitudes, in the apparent absence of candidate sunspot activity. It is also noteworthy that a disappearing filament appears to have been the most likely cause of one of the most major auroral storms of sunspot cycle 22, on the night of 1991 November 8–9 (section 2.8.1).

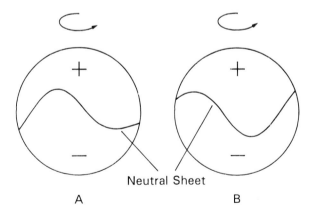

Fig. 5.5. Semi-permanent sectors of uniform overall magnetic polarity, separated by the solar wind's pleated neutral sheet (section 4.3.8) sweep across the Earth as the Sun rotates. Sector-boundary crossings, when the magnetic polarity of the solar wind flowing past the Earth reverses, are often marked by intervals of enhanced geomagnetic activity, associated with high-speed streams close to the 'folds' of the neutral sheet. In this schematic diagram, the folding of the neutral sheet has been exaggerated. At A, looking down the Earth-Sun line, a sector of negative magnetic polarity has just come into alignment; at B, some time later, this has rotated away to be replaced by a sector of positive magnetic polarity.

5.4.3 Magnetospheres of other planets

Spacecraft have explored the near environments of all the major planets with the exception of Pluto (which many astronomers contend is not a major planet in any case!). This exploration has revealed all to possess magnetic fields of differing intensities, which interact with the solar wind to varying extents. Studies of these offer important insights to the behaviour of the terrestrial magnetosphere.

Particles accelerated in their magnetospheres give rise to auroral displays in the atmospheres of the gas giants Jupiter, Saturn, Uranus and Neptune. These displays, on a vast scale, have been imaged on the planets' night-sides by the passing Voyager spacecraft.

5.4.3.1 Jupiter

Jupiter has by far the most extensive magnetosphere in the solar system, with a bow-shock lying 6 million kilometres upwind, and a magnetotail extending downwind beyond the orbit of Saturn (Van Allen, 1990). From the distance of Earth (4 AU at opposition), Jupiter's magnetosphere subtends an angle of about one degree (twice the apparent diameter of the Moon) against the sky surrounding the planet. The precise dimensions of the jovian magnetosphere are highly variable, ranging from 50–100 Jupiter-radii (3.6×10^6 to 7.2×10^6 km) depending on solar wind intensity.

Trapped plasma in Jupiter's magnetosphere is concentrated in an equatorial plane *magnetodisk* as a result of the system's rapid rotation. Measurements obtained during the pas-

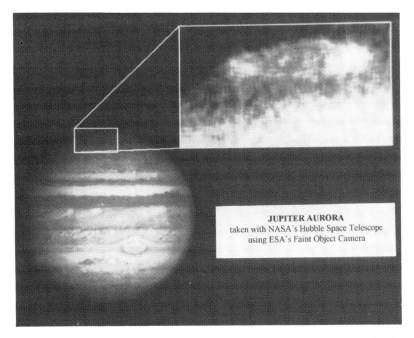

Fig. 5.6(a). The northern auroral oval of Jupiter, imaged using the Faint Object Camera aboard the Hubble Space Telescope. Under normal conditions, the jovian auroral ovals lie closer to the magnetic poles than those of the Earth. NASA photograph.

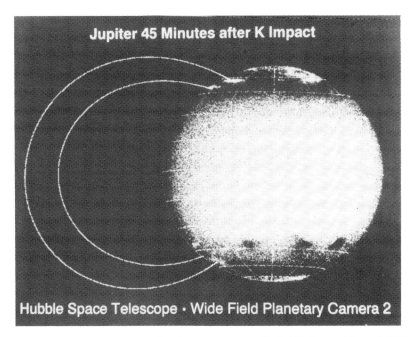

Fig. 5.6(b). Far-ultraviolet imaging of the jovian aurora using the Hubble Space Telescope's Wide Field Planetary Camera-2 on 19 July 1994 showed a marked enhancement of part of the northern oval 45 minutes following the impact of fragment K from the shattered comet Shoemaker-Levy 9. This activity is believed to be the result of energetic particles, ejected into the jovian magnetosphere during the impact, being delivered via magnetic field lines—as indicated—to the conjugate point in the opposite hemisphere. NASA photograph.

sage of the Ulysses spacecraft through near-Jupiter space in 1992 February indicate that the jovian radiation belts extend to a maximum latitude of 40° (compared with 70° for the terrestrial Van Allen belts). The jovian magnetospheric plasma population includes electrons from the planet's upper atmosphere and sulphur and oxygen ions (from sulphur dioxide) ejected during volcanic eruptions from the innermost Galilean satellite Io. The latter material forms an equatorial ring, the *Io torus*. Ulysses measurements showed the Io torus to be inhomogeneous, containing 'hot spots' of radio emission. The principal role of the solar wind appears to be in shaping, rather than populating, the jovian magnetosphere.

Particles in Jupiter's magnetosphere, by virtue of its immense size, can be accelerated to extremely high energies. Bursts of radio noise resulting from particle motions in the jovian magnetosphere are sufficiently powerful to be detected from Earth. These energetic particles present a hazard for spacecraft in the jovian system. Fortunately, although they sustained some damage as they traversed the near-Jupiter environment, the Pioneers and Voyagers survived to send back a wealth of information about Jupiter and, in the particular case of Voyager 2, the outer gas giants of the solar system.

The Voyagers imaged aurorae on Jupiter's night-side. Particles accelerated into Jupiter's atmosphere at high latitudes give rise to aurorae in ovals girdling either magnetic pole, as on Earth. The jovian aurorae, however, are 100 000 times as powerful as their terrestrial counterparts. Jupiter's aurorae appear predominantly red, thanks to the abundance of hydrogen in the planet's atmosphere.

More recently, observations of Jupiter's aurorae have been obtained on a routine basis using ultraviolet detectors aboard the IUE satellite, and ground-based infrared telescopes (Kim *et al.* 1991; Baron *et al.* 1991; Miller, 1992). Emissions of the H_3^+ ion, at a wavelength of 3.53 micrometres in the infrared have been of particular use in such studies. Around this wavelength, Jupiter appears dark as a result of atmospheric methane absorption. Jovian H_3^+ emissions can vary in intensity over timescales of an hour or so.

Together with other phenomena—principally the prominent large dark spots resulting from the impacts—the jovian aurorae were subject to intense scrutiny during the week of 1994 July 16–23, as the fragments of the disrupted Comet Shoemaker-Levy 9 crashed into the giant planet's atmosphere around 44°S latitude (Beatty and Levy, 1995).

Around the time of the impacts, 20-cm wavelength synchrotron radiation from electrons accelerated in the jovian magnetosphere was elevated by 25–50%, and remained at abnormally high levels for some months afterwards. Immediate effects related to the impacts included fading of sections of the southern auroral oval, and a brightening of the northern oval roughly 45 minutes following the impact of fragment K. The latter effect has been attributed to the arrival, along the connecting magnetic field line, of material from the equivalent (conjugate) point in the opposite hemisphere (Clark, 1995).

5.4.3.2   Saturn

Saturn's magnetosphere was first detected and measured by Pioneer 11 in 1979, and later examined in more detail by the Voyagers. As with Jupiter, Saturn has an extensive magnetosphere, on whose outer fringes orbits the large satellite Titan. Titan's atmosphere is a source of nitrogen which contributes to the particle population of the saturnian magnetosphere. The inner satellites may also contribute material, and appear to exert a complex control over the energy spectrum of electrons in the inner magnetosphere.

Saturn's magnetosphere contains a partial magnetodisk, and in some respects shows a structure intermediate between those of the Earth and Jupiter. As might be expected, the ring system has a role to play, in absorbing energetic particles: it appears that trapped, energetic particles are absent from the region of the magnetosphere bounded by the rings. Energetic particles rain into Saturn's atmosphere at high magnetic latitudes, there producing auroral activity.

5.4.3.3   Uranus and Neptune

Uranus is unusual in having an axis of rotation greatly tilted relative to the orbital plane. The magnetic axis of the planet is also considerably offset from its rotational axis (Hunt and Moore, 1989). Particle abundances measured by Voyager 2 during its close approach to Uranus in January 1986 indicate the magnetosphere to be populated chiefly by material from the planet's upper atmosphere (Miner, 1990). As with the saturnian system, the plasma population in Uranus' magnetosphere is controlled to some extent by the planet's satellites, and material in the tenuous ring system. Collisions between accelerated particles and the ring material may account for the darkness of the latter. Voyager detected auroral hydrogen emissions in Uranus' upper atmosphere.

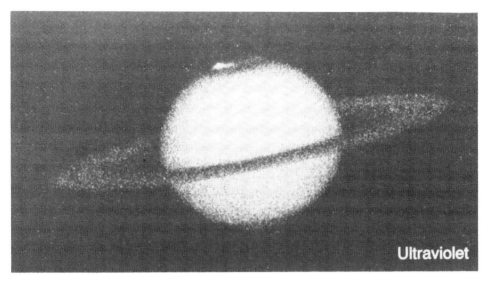

Fig. 5.7. The Wide Field Planetary Camera-2 aboard the Hubble Space telescope obtained this, the first image of Saturn's aurora, on October 9 1994. The saturnian aurora, emitting at wavelengths of 110–160 nm, is visualized in the far ultraviolet in this image. Changes in brightness and size were evident over timescales of as little as two hours. NASA photograph.

In a number of respects, Neptune—which Voyager 2 encountered in August 1989—is quite similar to Uranus. Neptune's axis of rotation is reasonably upright, but again, and surprisingly, its magnetic axis was found to be markedly offset. The neptunian magnetosphere is populated by particles from the planet's atmosphere, and from that of the main satellite Triton. Ionic species include hydrogen and nitrogen. Voyager detected weak auroral activity in Neptune's atmosphere (Mason, 1989), at an intensity much less than that of terrestrial aurora. Aurora is also generated in the atmosphere of Triton by particles accelerated in the neptunian magnetosphere (Hunt and Moore, 1994). Like those of Uranus, Neptune's thin rings are darkened by particle bombardment.

The offset of their magnetic axes has been taken to imply that the magnetic fields of Uranus and Neptune are generated in layers of those planets' interiors *above* their cores (Dowling, 1990). Much further investigation will undoubtedly be devoted to the mechanisms by which these fields are produced.

The magnetic fields of both Uranus and Neptune are also exceptional in being markedly tilted with respect to the planets' axes of rotation. Uranus' magnetic axis is tilted at 59° with respect to the axis, that of Neptune at 47° Both values are markedly greater than the tilts for any of the other planets.

5.4.3.4 Mercury

By virtue of its proximity to the Sun, Mercury experiences the most intense solar wind of all the planets. Despite its small size (4880 km diameter), Mercury has a relatively strong magnetic field, suggesting the existence of a proportionally large metallic core. The magnetic field of Mercury produces a bow-shock upwind of the planet. Mercury is large compared to its magnetosphere, and the planet's solid body fills the volume of space where trapping regions equivalent to Earth's Van Allen belts might otherwise form. Neither, in the absence of an appreciable atmosphere, does Mercury possess an ionosphere.

5.4.3.5   Venus and Mars

The two other terrestrial planets appear to have negligible magnetic fields and a consequent lack of magnetospheric development. Venus, in many respects regarded as Earth's twin despite its patently inimical atmosphere and searing temperatures, rotates only slowly on its axis (once every 243 Earth days). Venus is of a similar size to the Earth, and is expected to have a similar-sized core. A result of Venus' slow rotation, however, is the apparent absence of fluid motions in the core, and production of only a very weak magnetic field (1/25 000th that of the Earth).

Venus has a well-developed ionosphere resulting from the effects of solar radiation on its atmosphere. The ionosphere is involved with the principal interaction between Venus and the solar wind, creating a bow-shock about 2000 km (0.3 Venus-radii) upwind of the planet. Magnetic field lines in the solar wind flowing past Venus become draped around the planet and drawn into a long downwind 'tail' structure in a manner similar to those around comets (Russell, 1987).

At times of intense activity, the solar wind can be driven into the ionosphere of Venus, magnetizing it. Given the absence of a well-developed magnetosphere in which particles can be accelerated, in the near-Venus neighbourhood, auroral activity can be dismissed as the cause of the Ashen Light occasionally reported by visual observers of the planet.

Although its physical surface has been quite extensively studied and mapped by the Mariner and Viking probes, Mars has yet to be shown to possess any significant magnetic field. It is widely considered that, due to its relatively small size, Mars lacks a molten core in which the electric currents responsible for magnetic field generation can occur. As with Venus, the main solar wind interactions occur through the influence of the planet's ionosphere, creating a bow-shock upwind of Mars. Further exploration of the near-Mars environment is required, however, before definite conclusions can be drawn about the planet's magnetic status. In a way, it is rather sad to reflect that any future colonists of Mars will be unlikely to enjoy the ethereal beauty of the aurora in their skies!

5.4.3.6   The solar wind and small bodies

En route to Jupiter, the Galileo spacecraft was targeted close to two S-class asteroids, (951) Gaspra, and (243) Ida, encountered in 1991 October and 1994 August respectively (Beatty, 1993, 1995). S-class asteroids are believed to comprise silicate minerals with an admixture of iron. Both bodies are irregular in shape, and comparatively small: Gaspra has a long axis of 17 km, Ida 55 km. Magnetometer equipment aboard Galileo recorded considerable disturbances of the Interplanetary Magnetic Field as it passed both Gaspra and Ida, possibly indicative of a higher iron content than previously thought likely for such asteroids. Observations of this kind may be of value in the future determinations of asteroidal composition.

5.4.3.7   The zodiacal light

Although no longer considered a phenomenon of the Earth's high atmosphere, the diffuse glow of the *zodiacal light* (Bone, 1988), produced mainly by reflection of sunlight from myriads of small particles in the inner solar system does have some association with auroral effects. These particles, whose size is typically from 0.1 to 10 micrometres, are believed to originate principally from comets and asteroids. Estimates suggest that the inner solar system dust cloud contains about one cometary mass of these particles. Solar radiation pressure interactions (the Poynting–Robertson effect) result in a steady depletion of material from the dust cloud, such that without continued replenishment from new comets and asteroid fragment collisions it would disappear within a few thousands of years.

Observations obtained using photopolarimeters aboard the Pioneer spacecraft in the early 1970s indicated that the solar system dust cloud is shepherded by Jupiter's gravitational field such that the zodiacal light and *gegenschein* (counterglow) are absent outwith the giant planet's orbit.

The zodiacal light and associated skyglows are increasingly difficult to observe as light pollution spreads with urban growth. Few present-day astronomers have seen the zodiacal light, while the *gegenschein* is even more elusive. These glows were, however, known to astronomers at least as far back as the seventeenth century when Cassini carried out systematic observations of the zodiacal light, and noted that it varied in brightness from time to time.

The zodiacal light's brightness variations may be correlated with the solar cycle. It has been suggested that the zodiacal light is brighter at sunspot minimum than at sunspot maximum. A probable explanation for this is the presence of energetic coronal hole particle streams (section 4.3.5) permeating the solar system in the years around sunspot minimum, producing excitation of the tenuous interplanetary medium. Brightening of the zodiacal light on short timescales of a few days has also been reported by some observers, and may be connected to solar flare events. One such brightening, reported in 1990, came a few days after a fairly vigorous auroral display penetrating to the latitudes of the English Midlands (Graham, 1990).

Short-term variability in the intensity of the zodiacal light may also be accounted for by cometary activity. For example, several bright Sun-grazing comets were detected by Solar Max and the Solwind instrument, having been missed by more conventional patrols; disruption of such comets on close approach to the Sun could lead to transient increases in reflective dust in the zodiacal cloud.

Typically, the zodiacal light appears as a broad cone of diffuse light, little brighter than the Milky Way, and is best seen, from temperate latitudes, either in the evening sky about 90 minutes after sunset around the spring equinox, or in the morning sky around the autumnal equinox. At other times, it cuts a shallow angle relative to the horizon and is lost among the haze. The zodiacal light is seen to best advantage from the tropics, where the angle of the ecliptic to the horizon is steep. The faint *gegenschein* is only visible in the very darkest of skies (Meinel and Meinel, 1991), appearing as a diffuse oval of reflected sunlight some 10° or 20° in diameter in the midnight sky directly opposite the Sun.

REFERENCES

Akasofu, S.-I. (1982) The aurora: new light on an old subject. *Sky and Telescope* **64** 534–537.

Akasofu, S.-I. (1989) The dynamic aurora. *Scient. Am.* **260** (5) 54–63.

Akasofu, S.-I., and Kamide, Y. (1987) The aurora. In: Akasofu, S.-I., and Kamide, Y.

(Eds), *The solar wind and the Earth*. D. Reidel.

Allen, J. B. (1993) Satellite finds new radiation belt. *Astronomy* **21** (11) 26.

Anderson, D. L. (1990) Planet Earth. In: Beatty, J. K., and Chaikin, A. (Eds), *The new solar system* (3rd Edn). Cambridge University Press.

Baron, R., Jospeh, R. D., Owen, T., Tennyson, J., Miller, S., and Ballester, G.E. (1991) Imaging Jupiter's aurorae from H_3^+ emissions in the 3–4μm band. *Nature* **353** 539–542.

Beatty, J. K. (1993) The long road to Jupiter. *Sky and Telescope* **85** (4) 18–21.

Beatty, J. K. (1995) Ida & company. *Sky and Telescope* **89** (1) 20–23.

Beatty, J.K., and Levy, D. H. (1995) Crashes to ashes: a comet's demise. *Sky and Telescope* **90** 4 18–26.

Bloxham, J., and Gubbins, D. (1989) The evolution of the Earth's magnetic field. *Scient Am.* **261** 6 30–37.

Bond, P. (1993) Close encounter with a comet. *Astronomy* **21** (11) 42–47.

Bone, N. (1988) The zodiacal light. *Astronomy Now* **2** (9) 17–21.

Brandt, J. C. (1990a) Comets. In: Beatty, J. K., and Chaikin, A. (Eds), *The new solar system* (3rd edn). Cambridge University Press.

Brandt, J. C. (1990b) The large-scale plasma structure of Halley's comet, 1985–1986. In: Mason, J. W. (Ed.), *Comet Halley,* Vol. 1, *Investigations, results, interpretations*. Ellis Horwood.

Clark, S. (1995) Blow by blow: the death of a comet. *Astronomy Now* **9** (1) 39–44.

Coates, A., and Mason, K. (1992) Cometary tails. *Astronomy Now* **6** (7) 47–49.

Dowling, T. (1990) Big, blue: the twin worlds of Uranus and Neptune. *Astronomy* **18** (10) 42–53.

Graham, D. (1990) Junior Astronomical Society Circular 155.

Hones, E. W. (1986) The Earth's magnetotail. *Scient. Am.* **254** (2) 32–39.

Hunt, G., and Moore, P. (1989) *Atlas of Uranus*. Cambridge University Press.

Hunt, G. E., and Moore, P. (1994) *Atlas of Neptune*. Cambridge University Press.

Intriligator, D. S. and Dryer, M. (1991) A kick from the solar wind as the cause of comet Halley's February 1991 flare. *Nature* **353** 407–409.

Jastrow, R. (1959) Artificial satellites and the Earth's Atmosphere. *Scient. Am.* **201** (2) 37–43.

Jeanloz, R. (1983) The Earth's core. *Scient. Am.* **249** (3) 40–49.

Kim, S. J., Drossart, P., Caldwell, J., Maillard, J.-P., Herbst, T., and Shure, M. (1991) Images of aurorae on Jupiter from H_3^+ emission at 4μm. *Nature* **353** 536–539.

Lockwood, M., Denig, W. F., Farmer, A. D., Davda, V. N., Cowley, S. W. H., and Luhr, H. (1993) Ionospheric signatures of pulsed reconnection at the Earth's magnetopause. *Nature* **361** 424–428.

Mason, J. (1989) Encounter with the blue giant. *Astronomy Now* **3** (10) 27–30.

Meinel, A., and Meinel, M. (1991) *Sunsets, twilights and evening skies*. Cambridge University Press.

Miller, S. (1992) Bright lights on Jupiter. *Astronomy Now* **6** (3) 46–48.

Miner, E. D (1990) *Uranus: the planet, rings and satellites*. Ellis Horwood.

Powell, C. S. (1991) Peering inward. *Scient Am.* **264** (6) 72–81.

Ratcliffe, J. A. (1972) *An introduction to the ionosphere and magnetosphere*. Cambridge University Press.

Russell, C. T. (1987) The magnetosphere. In: Akasofu, S.-I., and Kamide, Y., (Eds), *The solar wind and the Earth.* D. Reidel.

Van Allen, J. A. (1990) Magnetospheres, cosmic rays, and the interplanetary medium. In: Beatty, J. K., and Chaikin, A., (Eds), *The new Solar System* (3rd edn). Cambridge University Press.

6

The polar aurora

6.1 COMPOSITION AND NATURE OF THE HIGH ATMOSPHERE

Some early observers believed that auroral forms could, on occasion, be seen at heights similar to those of nearby mountains. Accurate photographic triangulation of the aurora by Stormer and others has long since dispelled this notion. Likely causes of the illusion, of aurora 'dancing' around mountain peaks, include arctic mirages produced in cold air masses. The visible aurora occurs mainly at heights in excess of 100 km, and is therefore strictly a phenomenon of Earth's high atmosphere. In any consideration of the aurora, it is worth examining the milieu in which it takes place, since this has a fundamental bearing on the nature of activity.

6.1.1 Atomic and molecular composition; density

The atmosphere at 100 km altitude, the base of the auroral layer, is very tenuous, having a pressure similar to that within a domestic lightbulb. Pressure continues to decrease with increasing altitude (Fig. 6.1), and that at 300 km, the altitude of higher forms within auroral draperies, is equivalent to a high-quality laboratory vacuum (Mitra, 1952). Particle densities fall from 2.5×10^{19} cm^{-3} at ground level to 3.7×10^{13} cm^{-3} at 100 km altitude, and only 7.3×10^{5} cm^{-3} at the top of the atmosphere, where the highest auroral forms are seen, around 1000 km altitude.

The principal atmospheric constituents involved in auroral emissions are nitrogen and oxygen, excited by energetic electrons from the neutral sheet of the magnetosphere (section 5.2), accelerated into the high atmosphere in response to fluctuations in solar activity, transmitted to the Earth via the solar wind (section 4.3.8). Oxygen and nitrogen are, of course, the main constituents of the atmosphere, but have a differing abundance ratio and nature at the height of the aurora from the familiar 78% N$_2$: 21% O$_2$ observed in the troposphere.

Above about 80 km altitude, photodissociation by ultraviolet radiation from the Sun converts O$_2$ molecules to atomic oxygen, OI:

$$O_2 \xrightarrow{h\upsilon} O + O$$

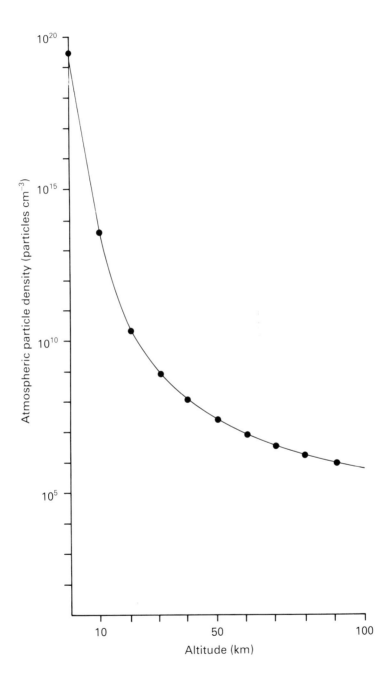

Fig. 6.1. Particle density as a function of height in Earth's atmosphere. At auroral heights, the atmospheric density is equivalent to a good-quality laboratory vacuum.

The atomic oxygen concentration reaches a maximum at about 105 km. Relative proportions of gas species at 100 km are 76.5% N_2 : 20.5% O_2 : 3.0% OI. Above this height, atomic oxygen becomes the predominant species, being approximately 100-fold more abundant than N_2 at 400 km altitude, while molecular oxygen is effectively absent above 130 km.

Ionized nitrogen, N_2^+, is also found in the high atmosphere, and is another product of the absorption of solar radiation, in the extreme ultraviolet at wavelengths shorter than 79.6 nm:

$$N_2 \xrightarrow{h\upsilon} N_2^+ + e^-$$

N_2^+ may also be produced as a result of collisions between electrons and molecular nitrogen:

$$N_2 + e^- \rightarrow N_2^+ + 2e^-$$

Dissociative recombination of N_2^+ following collisions with energetic electrons produces nitrogen atoms (NI) in the regions where auroral activity occurs:

$$N_2^+ + e^- \rightarrow N + N$$

Atomic oxygen, and molecular and atomic nitrogen are, therefore, the principal atmospheric species available for, and involved in, the production of auroral emissions.

6.1.2 The ionosphere

An important effect of the action of sunlight on the Earth's upper atmosphere is photodissociation (section 6.1.1), and consequent *ionization*, of atmospheric constituents. The chief components of sunlight responsible for this ionization are short wavelength emissions in the X-ray and ultraviolet regions of the spectrum. This ionization produces atmospheric populations of positively charged ions and negatively charged electrons.

Ionization appears at a number of atmospheric levels, producing layers which may be identified by their interactions with radio waves. Collectively, the high atmosphere layers of ionization at heights above 60 km or so are referred to as the *ionosphere*. Investigations since the 1920s have revealed much about the structure, nature, and temporal behaviour of the ionosphere (Ratcliffe, 1970).

The first ionospheric layer to be recognized was the *Heaviside layer*, now more commonly known as the *E-layer*. The E-layer is used by radio operators as a reflective surface from which signals can be 'bounced' to distant stations lying over the horizon, and beyond range of horizontally transmitted ground waves. Short-wave radio reflection from the ionospheric E-layer shows a diurnal variation, being enhanced during the hours of darkness. This effect is a consequence of the increased *absorption* of radio waves during daytime, when sunlight produces higher electron densities.

The E-layer lies at an altitude of about 110 km (similar to the base atmospheric level of the visible aurora), and is present over the whole globe.

Early studies of the ionosphere soon suggested the existence of further layers. The E-layer is transparent to very high frequency radio waves, and from reflection of these Appleton was able to show the existence of a higher ionization region in the atmosphere, the

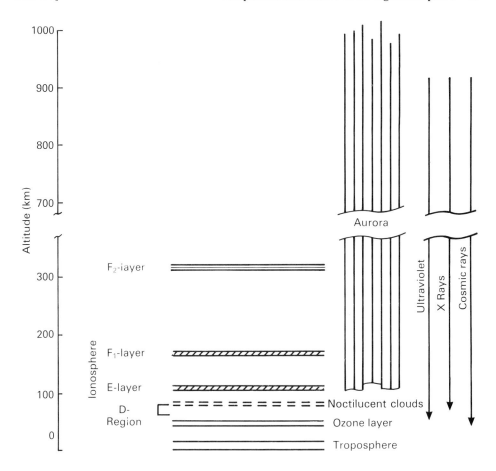

Fig. 6.2. Broad cross-section through the Earth's atmosphere, showing its structure and some of the relevant phenomena which occur there, including the aurora. Auroral activity typically extends from the ionospheric E-layer to altitudes in excess of 1000 km. Occasionally, intense activity resulting from extremely high energy particles can be seen at lower altitudes, down to perhaps 65 km (Table 6.2). Also shown are the maximum penetration levels of ultraviolet and X-ray emissions from the Sun, and of primary cosmic ray particles.

F-layer. The F-layer shows a partial split into two regions, a lower F_1 region (at about 160 km altitude), and upper F_2 region (300 km). Diurnal effects again operate. The F_2 layer disappears at night, and may also disappear during daytime on occasion.

The F-layer, like the E-layer, reflects radio waves. The lowest region of the ionosphere, the D-region between 65 and 80 km altitude, however, principally *absorbs* radio waves. D-region absorptions occur in response to solar activity, and are frequently noted at the time of sunspot maximum. Both ultraviolet and X-rays from the Sun produce ionization in the upper D-region. The X-ray flux is more variable, and is also the major source of variation in D-region ionization. Increased ionization of the D-layer, produced by the enhanced emission of X-rays during solar flares, gives rise to *Sudden Ionospheric Disturbance* (SID) events. SIDs affect only the day-side D-region, and bring about abrupt increases in radio absorption. These events may last a matter of minutes, up to an hour or so. Ionization produced during SIDs is fairly rapidly lost as electrons recombine with positive ions in the relatively dense D-region of the atmosphere.

Atmospheric absorption prevents penetration of X-rays below about 60 km, while the stratospheric ozone layer—fortunately for living organisms at the Earth's surface—cuts off ultraviolet penetration below around 50 km altitude. Ionization in the lower parts of the D-region is caused by the penetration of energetic cosmic rays. An apparent paradox, resolved on development of models of the solar wind and heliosphere (section 4.3.9), was the observation that lower-level D-region ionization reaches a minimum at sunspot maximum, when upper D-region ionization peaks. It is now known that the flux of high-energy cosmic rays reaching the inner solar system is reduced by the increased heliospheric magnetic field intensity around sunspot maximum (section 8.3).

Polar cap absorptions, again in the D-region, are an important cause of short-wave radio blackouts at high latitudes. These result from an increase in ionization of the middle atmosphere by highly penetrative protons ejected during solar flare events, and are closely associated with auroral phenomena (section 6.4.2).

Auroral activity can have a number of effects on the ionosphere, leading to disruption of short-wave radio communication. There are times, however, when auroral conditions in the ionosphere are *beneficial* to radio communication, allowing longer than normal distance contacts to be made (section 8.4).

Photodissociation of atmospheric components by sunlight is, obviously, a daytime phenomenon. The mechanisms by which ionization is lost are important in shaping the ionosphere. In the absence of continued solar excitation, many of these loss processes operate at night, involving recombination between electrons and positive ions. Some of these recombination processes release the diffuse, weak background light of the airglow (section 9.3).

6.2 AURORAL EMISSIONS; SPECTROSCOPY AND ATOMIC STRUCTURE

Auroral light results largely from the excitation of atmospheric atoms and molecules during collisions with energetic particles, and the subsequent re-emission of the imparted excess energy in the form of light. Spectroscopic studies of the aurora have yielded much information about the nature of these emissions, and the particles involved. Spectra recorded using plane diffraction gratings have been analysed in detail. The principal atomic species involved in auroral emissions are oxygen, nitrogen and hydrogen (Vallance Jones, 1991).

The light of the visual aurora is produced almost entirely by collisions between electrons and atmospheric oxygen and nitrogen. The energies of incoming accelerated particles are indicated in Table 6.1 (after Livesey (1989): by kind permission of the author). Table 6.2 highlights some of the important visually observed auroral emissions.

For a given chemical species, there is a series of permitted energy levels surrounding the positively charged nucleus in which electrons may orbit. These can be predicted from the laws of quantum mechanics (Hey and Walters, 1987). Under normal circumstances, electrons sequentially fill the lowest energy levels (closest to the nucleus) first. Electrons may, at the expense of appropriate quanta of energy, be transferred between levels—that is, undergo *transition*. Transitions to higher energy (outer) orbital levels require the addition of energy from outside: in the case of atmospheric oxygen and nitrogen during auroral conditions, this is imparted by incoming electrons or protons accelerated in the magnetosphere, or from other atmospheric particles which have themselves previously undergone *excitation*.

Table 6.1. Penetration of energetic particles into Earth's atmosphere

Particle	Energy (keV)	Penetration (km altitude)
Electron	1	150–200
	10	100
	30	90
Proton	30 000	50
	500 000	Ground level during severe polar cap event

Table 6.2. Some visible auroral emissions

Wavelength (nm)	Emitting species	Typical altitude (km)	Visual colour
391.4[a]	N_2^+	1000	Violet-purple
427.8	N_2^+	1000	Violet-purple
557.7	OI	90–150	Green
630.0	OI	>150	Red
636.4	OI	>150	Red
656.3	Hydrogen-alpha	120	Red
661.1	N_2–1P	65–90	Red
669.6	N_2–1P	65–90	Red
676.8	N_2–1P	65–90	Red
686.1	N_2–1P	65–90	Red

[a] Also at 50-70 km altitude during Polar Cap Absorptions (section 6.4.2).

Sufficient energy may be delivered to produce electron transitions to levels not normally occupied in these atmospheric species. Such energy levels cannot usually be occupied since the presence of electrons therein violates the quantum mechanical rules which demand that lower levels must be filled first. These transitions are therefore described as 'forbidden'. Emissions resulting from forbidden transitions are seen only in conditions of rarefied particle density, such as obtain at high altitudes in Earth's atmosphere (section 6.1.1).

Excited species are *metastable*: after a given time interval, the excitation is lost as the electron drops back to a lower orbit, and the atmospheric particle returns to its preferred minimal-energy *ground state*. The excess energy is re-emitted as a photon of light whose wavelength is precisely governed by the same quantum rules which dictate the available energy levels surrounding a given nucleus. Figs 6.3 and 6.4 present some of the predicted (and observed!) spectral emissions associated with transitions between excited and ground states in atmospheric oxygen and nitrogen during auroral activity.

The presence of Doppler-shifted Balmer lines in auroral spectra indicates the involvement of protons (*hydrogen* nuclei) in producing some of the excitation which results in auroral emissions (Vallance Jones, 1969). Hydrogen emissions are commonly detected at altitudes around 120 km.

Visually, the aurora may often appear white or colourless when faint. Bright displays can show marked colour, however, notably greens and reds. The dominant feature of the visible auroral spectrum is the forbidden green emission at 557.7 nm, resulting from the excitation of atomic *oxygen* (OI). This characteristic 'auroral green line' may be detected using narrow-passband interference filters, allowing the observation of activity in moonlit or light-polluted skies, or in cloudy conditions.

Green atomic oxygen emission is dominant in the lower parts of auroral displays, around 100 km altitude. Higher in the atmosphere, collision of lower-energy electrons with atomic oxygen gives rise to red emissions at 630.0 nm and 636.4 nm wavelengths.

Atomic oxygen has two principal excited states which give rise to visual auroral emission. The 557.7 nm emission results from the second, higher excited state, which requires a greater energy input. The lifetime of OI in the second excited state is short (0.74 second), and electrons in this energy level rapidly decay to the lower level of the first excited state, yielding photons at 557.7 nm wavelength in the process (Fig. 6.3). Red oxygen emissions result from the decay of electrons from the first excited state to ground state.

The lifetime of the first excited state of OI is longer (110 seconds) than that in the second excited state, leading to the observed variations in the predominant emission with height. In the lower, denser parts of the auroral layer, the slow red oxygen emissions are *quenched*: the time required for an excited electron to return to ground state is greater than the average interval between collisions of oxygen with other atmospheric particles. The excitation energy is therefore lost through collisions before it can be re-emitted. Conversely, the green auroral emission occurs rapidly, before most excited oxygen atoms can give up their excess energy through collision.

The production of two red emission lines following the decay of electrons from the first excited state in atomic oxygen results from the availability of two, almost equivalent, energy levels in the normal outer quantum shell around the nucleus, either of which may require filling to restore the ground state configuration.

At altitudes of 1000 km, where the uppermost parts of auroral rays may reach above the Earth's shadow at certain times of year, molecular *nitrogen* is substantially ionized to N_2^+ by the action of solar ultraviolet (section 6.1.1): approximately 75% is in this form at the top of the auroral layer, compared with 20% of atmospheric nitrogen at 100 km altitude. Excitation of N_2^+ by incoming accelerated electrons produces blue-purple emissions at 391.4 nm and 427.8 nm wavelengths. N_2^+ emissions are also present in auroral features at altitudes around 100 km; the 391.4 nm emission may occur as low as 50–70 km during

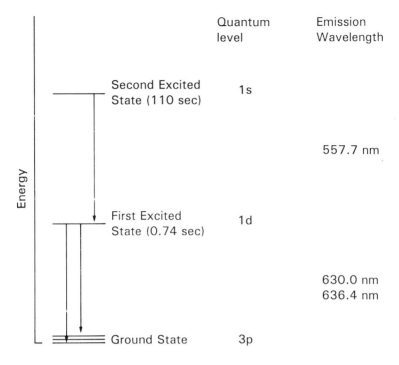

Fig. 6.3. Energy level diagram for atomic oxygen, showing the electron transitions commonly found involved in aurorae, and the light emission wavelengths produced when excited electrons return to lower energy states. Lifetimes in the excited states are also shown.

polar glow aurora (section 6.4.2). In sunlit aurora, however, the blue-purple emissions are enhanced by absorption and re-emission of these wavelengths from solar radiation, a process of *resonance* which results in the 391.4 nm and 427.8 nm lines being stronger than they appear when produced by the collisional mechanism alone.

Red auroral emissions seen on the undersides of arcs or bands in the most vigorous aurorae are produced by very energetic (30 keV) electrons penetrating to lower levels—of the order of 90 km altitude—and exciting molecular nitrogen (N_2). Transitions to the first positive state (N_2–1P; Fig. 6.3) produce these emissions in a group of four spectral lines between 661.1 and 686.1 nm wavelength.

All these emissions can, at times be so weak that the visual observer is unaware of their occurrence. Time-exposure photography using fast films and narrow-passband interference filters matched to auroral emissions can be used to determine the extent of low-energy auroral emission on a given night (Simmons, 1985).

The spectrum of the aurora, and the processes which give rise to it, are complex. Balloon, rocket and satellite measurements have extended spectral studies into the ultraviolet and infrared regions of the spectrum, in which such as the N_2 Vegard–Kaplan and N_2^+ Meinel emission are produced respectively. Emissions from atomic nitrogen (NI) also appear in these regions. Detectors aboard Dynamics Explorer-1 recorded an ultraviolet emission of atomic oxygen at 130.4 nm in order to produce unparalleled images of the auroral ovals taken from above the poles (section 3.9).

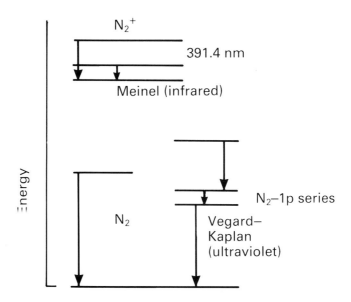

Fig. 6.4. Energy level diagram for molecular nitrogen (N₂), and ionized molecular nitrogen (N₂⁺). The 391.4 nm emission is enhanced by resonance scattering in sunlit aurorae, and appears bluish-purple to the eye. Sunlit aurorae are often seen at high latitudes during late spring or early autumn.

6.3 GEOGRAPHICAL DISTRIBUTION OF AURORAE

It was surmised during the nineteenth century that aurorae are most frequently seen in geographical zones of latitude, 10° wide and centred around 65° latitude (section 3.4). Aurorae are seen more often as one travels polewards towards the auroral zones from temperate latitudes. The frequency with which aurorae are seen actually falls off again, however, with increasing latitude polewards from the auroral zones, a fact not always appreciated by writers of popular astronomical texts.

6.3.1 The auroral ovals

Scientific measurements of the location of auroral features suggested, by the late 1950s, that auroral activity at any one instant occurred principally within oval regions disposed around the geomagnetic poles in either hemisphere. The existence of these *auroral ovals* was confirmed by observations from satellites in high-altitude polar orbits. The Dynamics Explorer-1 satellite (section 3.10) has been of particular value in studies of the behaviour of the auroral ovals.

The northern and southern hemisphere auroral ovals are direct mirror-images of one another, as has been confirmed by simultaneous photography of visible auroral forms from NASA aircraft flying at conjugate points (locations of equivalent magnetic latitude and

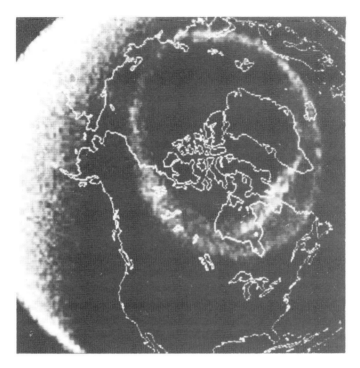

Fig. 6.5. A Dynamics Explorer-1 view of the northern hemisphere auroral oval taken from high orbit. The outline of continental North America has been added to indicate the geographical extent of the aurora. A substorm is in progress, with a bright band of aurora over northeast Canada indicating the westwards-travelling surge. University of Iowa photograph.

longitude, essentially at opposite ends of magnetic field lines looping out from the Earth) in either hemisphere. Activity in the northern oval is duplicated in the southern.

Relative to the geomagnetic poles, the auroral ovals are displaced, such that they extend further towards the equator at the magnetic midnight point (in a straight line through the pole to the Sun) on the night-side than on the day-side. The regions of the Earth which enjoy the highest incidence of visible aurora displays are those lying reasonably close to the greatest equatorwards extent of the auroral ovals under undisturbed conditions. In the northern hemisphere, this *auroral zone* lies in a ring stretching from the North Cape of Norway, to Iceland, northern Canada, Alaska, and on to the northern Russian states. The observer is carried towards this point on the auroral oval around magnetic midnight each day by the Earth's rotation. The auroral ovals can be regarded, under quiet conditions, as fixed in space above the rotating Earth.

The equatorwards displacement of the auroral oval on the night-side is a direct reflection of the distortion of the terrestrial magnetic field by the influence of the solar wind, in common with the drawing-out of the magnetosphere into its 'wind-sock' configuration (section 5.2). Effectively, the auroral oval is the ionospheric link in the complex circuit flowing around the magnetosphere. When the power of the circuit increases in response to a southwards-directed IMF, the ovals increase in their geographical extent, particularly, again, on the night-side, and the active aurora becomes visible from lower latitudes (section 7.2).

Under quiet conditions, the auroral ovals form rings of 4000–5000 km diameter around the geomagnetic poles. Their extent can become greatly increased during major geomagnetic storms. During the Great Aurora of March 1989, the northern hemisphere oval's southern limit extended to the latitudes of the Iberian peninsula and the mid-Atlantic, as imaged from Dynamics Explorer-1.

| Evening | Midnight | Morning |

Fig. 6.6. An observer at a given location ('X' in the diagram) is carried closest to the position of the quiet-condition auroral oval by the Earth's rotation around magnetic midnight, being less favourably placed in early evening and early morning. The auroral ovals can be regarded as more or less fixed in space above the rotating Earth: the maximum zones of auroral frequency are those which are carried nightly under the normal, undisturbed maximum equatorwards extent of the ovals.

6.3.2 Stable Auroral Red (SAR) arcs

Distinct from the polar oval aurorae, and lying in middle latitudes even at times of relatively low geomagnetic activity, are the Stable Auroral Red (SAR) arcs. These form high in the atmosphere, between altitudes of 300 and 700 km, reaching at their lowest extent into the F-layer of the ionosphere. Their weak (typically sub-visual) emissions are produced by atomic oxygen excitation at 630.0 and 636.4 nm wavelengths, and result from penetration by the earthwards edge of the magnetotail plasma sheet into the plasmasphere (section 5.2).

SAR arcs brighten in response to enhanced geomagnetic activity. The arcs have an east—west extent of several thousands of kilometres, and span up to 600 km (5°) in latitude.

6.4 THE APPEARANCE OF AURORAL ACTIVITY AT HIGH LATITUDES

Aurorae are, in the popular imagination, almost synonymous with the polar regions. Countless books have appeared over the years covering the exploration of the Arctic and Antarctic and, latterly, on their wildlife. The aurora receives frequent mention and illustration in such texts. In the writings of the late-nineteenth century Norwegian explorer Nansen (section 2.5), clear descriptions of auroral substorm activity can be found, for example. Notably, when at very high latitudes in the Arctic, Nansen saw auroral activity predominantly to the south (equatorwards) under quiet conditions, and a migration north-wards during the active phase of the westward travelling surge (section 5.4.2.1).

6.4.1 Polar cusp aurorae

Most of the solar wind material approaching the Earth passes around the bow-shock and on into interplanetary space without any magnetospheric interaction. A certain, very im-portant, proportion of solar wind plasma can, however, enter near-Earth space via the high-latitude polar cusp regions (section 5.4.2) on the day-side of the auroral oval. Elec-trons trickling into the high atmosphere at these locations give rise to pallid, weak auroral displays, which may cover the entire sky with a diffuse background glow. Polar cusp auro-rae are seen only at high latitudes during midwinter, when the noon sky is still sufficiently dark to render auroral activity visible. (By local midnight, the Earth's rotation has carried the high-latitude observer away from under the polar cusp). One such location is Advent-dalen (78°17′N, 15°55′E) on Spitsbergen Island in the Barents Sea, the site of a profes-sional auroral observatory, the Nordlysstasjonen operated by the University of Tromso. At Adventdalen, a 6-week observing season, centred on the winter solstice, favours the recording of polar cusp aurora.

Dayside aurorae seen under undisturbed conditions in the midwinter noon at high lati-tudes follow a regular pattern (Simmons and Henriksen, 1994). For a period of 1–4 hours before local noon, diffuse patchy green aurora, with emission predominantly at 557.7 nm, is seen, resulting from excitation of atomic oxygen by incoming electrons with energies below 500 eV arriving via the polar cleft. For the 2-hour period around noon, diffuse weak (often sub-visual) red OI emissions at 630.0 and 636.4 nm, produced by low-energy (10–50 eV) electrons entering from the solar wind directly via the polar cusp, are seen. Cleft electrons then give rise to discrete green (557.7 nm emission) arcs for an interval of 1–4 hours following noon.

6.4.2 Polar Cap Absorptions (PCA)

Violent events, associated with flares, in the inner solar atmosphere can accelerate protons to energies of between 1 and 100 MeV. High-energy protons may arrive in near-Earth space via the solar wind soon (6–10 hours) after the observed flare, and are guided by the magnetic field into the polar cap regions within the auroral ovals. Here, ionization of at-mospheric particles by energetic protons gives rise to polar cap absorption (PCA) events, which can last for several days. Increased electron density in the D-region of the iono-sphere at 80 km altitude results in absorption of high-frequency radio waves within the auroral zones at such times. The occurrence of PCAs appears to correlate with those solar flares which produce type IV radio bursts (Hultqvist, 1969; section 4.3.6).

Associated with PCA events are diffuse, weak (usually sub-visual) *polar glow aurorae*, which may fill most or all of the polar cap region. These events are most readily studied using photometry or spectroscopy (Lassen, 1969). When sufficiently bright to be notice-able to the naked eye, polar glow aurora appears pink or red as a result of enhanced N_2–1P emissions and OI emissions at 630.0 and 636.4 nm. N_2^+ 391.4 nm emission also shows enhancement. Riometric studies indicate these emissions to occur at altitudes of 50–70 km (much lower than most auroral features), equivalent to the expected penetration of protons in the 5–30 MeV range. Surprisingly, given the involvement of protons in these events, hydrogen emissions are weak in polar glow aurorae.

Enhanced N_2^+ 391.4 nm emission occurs as soon as accelerated protons begin to arrive, reaching a peak some 24 hours later—coincident with the arrival of the main shock-wave produced in the solar wind by the flare-associated event, which brings the onset of magnetic SSC about 24 hours later (Elvey, 1965; Sandford, 1967).

Photographic studies, aided by narrow-passband interference filters, of polar glow aurora have been carried out from Spitsbergen (Simmons and Henriksen, 1995). The days around the major auroral storm of 1986 February 8–9 (section 2.6) were accompanied by visible polar cap aurora, possibly showing striation; such displays are more usually reported as homogeneous.

6.4.3 Substorm and geomagnetic storm events

Under normal conditions, aurorae are most frequently observed at high latitudes between about 17 hours and 02 hours local time. This is the period during which the Earth's rotation carries the high-latitude observer closest to the auroral oval's maximum equatorwards extent. Also, the most intense auroral activity occurs in the pre-midnight sector of the oval.

As found by Akasofu and others following studies of the International Geophysical Year data, and in accordance with the subsequent Dynamics Explorer and other satellite observations, there are marked differences between the quiet condition aurora observed in the evening and morning sectors of the auroral oval. The evening sector is dominated by greenish arcs, bands or rayed draperies. In contrast, the forms seen in the morning sector tend to be reddish and more fragmentary and diffuse (Simmons, 1988).

Observers close to the maximum zone of auroral activity swept out daily by the quiet-time auroral oval's greatest equatorwards extent (section 6.3.1) are well positioned to witness the occurrence of *substorm* activity (section 5.4.2.1). The pattern seen at these high-latitude sites on a given night will depend on a number of factors. These include the precise time at which the disturbance leading to the substorm reaches its peak, and the overall level of geomagnetic activity.

Commonly, auroral activity at high latitudes will begin in the evening as quiet *arcs*, often multiple, and usually homogeneous. The onset of increased activity may be heralded by brightening of the arcs, and the appearance of vertical ray structures. During the expansive phase, rayed arcs and folded *band* forms may rapidly extend across the whole sky, showing strong colour (green rayed structures with a red lower border resulting from N_2 emission (section 6.2), for example) and vigorous movement. This increase in north–south extent is a consequence of overall broadening of the night-side auroral oval. Loop or spiral structures may appear as the the auroral oval is contorted by the westward-travelling surge. Eastwards counterstreaming may also be seen in the rays. The narrowness of individual bands or arcs becomes apparent as these pass overhead; while rayed structures may have a vertical extent of several hundreds of kilometres, and an east–west extent of thousands of kilometres, each may be only 1 km or so in width. Folded bands, as they pass overhead, will show coronal structure as a result of perspective. This activity, lasting for perhaps 30 minutes, is described as 'auroral breakup' (Davis, 1992).

Fig. 6.7. Folding of auroral bands, associated with the westwards-travelling surge during substorms, produces 'curled' structures such as these, as seen from high latitudes.

Auroral breakup is particularly spectacular if it occurs around local midnight: the equatorwards edge of the auroral oval at its midnight point undergoes the most marked brightening. When breakup occurs in the early evening, the observer may find much of the activity in the eastern part of the sky.

As activity subsides, the auroral forms become more diffuse and fade, taking on the appearance of 'old rays' as described by Gartlein when attempting to categorize auroral forms in the 1950s. Such aurora is common in the morning sector of the oval. These pulsating forms mark the recovery phase of the substorm. Even at times of low activity, following substorms, the sky may be suffused by a faint background of auroral light.

Following breakup, and the recovery phase, quiescent arcs may re-form, prior to the onset of a further round of activity. Under intensely disturbed geomagnetic conditions resulting from violent sunspot activity, substorms may follow closely one upon an other, leading to hugely impressive aurorae maintained through much of the night. On the other hand, when geomagnetic activity is low, substorms may comprise little more than a brightening of the arc structures in the evening sector, with no ensuing major breakup.

Several amateur enthusiasts in the United States have made trips to high latitudes in Alaska or Canada, specifically to photograph the aurora. Guidelines on auroral photography are given later (section 7.2.6): in general, the required exposure times during auroral breakup are rather shorter than those for displays at lower latitudes. Bright, rapidly moving forms may be captured on exposures of as little as 1–2 seconds on ISO 400 film at $f/2$.

Major geomagnetic storm activity can also carry the aurora equatorwards *away* from high-latitude skies, however, although the initial stages may show an activity pattern similar to the onset of substorms, including brightening of forms and development of the west-ward-travelling surge. It is documented, for example, that during the great mid-latitude storm of January 25 1938, observers in southern Europe were able to see red aurora across much of the sky, while those in Tromso, Norway saw no aurora whatsoever (Moore, 1979)!

6.4.4 Theta aurora

One of the surprising discoveries made by the Dynamics Explorer-1 satellite was the oc-currence of bridges of weak auroral activity connecting across the inside of the auroral oval, and aligned to the noon–midnight line (Fig. 6.8). These transpolar arcs, commonly described as 'theta aurora', from the resemblance of the auroral oval to the Greek letter when they are present, appear at times when the Interplanetary Magnetic Field turns north-wards, magnetospheric activity in general declines, and the auroral oval fades. Theta au-rora events are accompanied by a pronounced weakening of the auroral electrojets.

Under low-activity conditions, an observer at a high-latitude site, within the auroral oval, is carried under the theta aurora twice daily—around noon and midnight—by Earth's rotation. The north–south alignment of transpolar arc features contrasts with the normal east–west alignment of substorm or storm arcs in the auroral oval population. Studies of

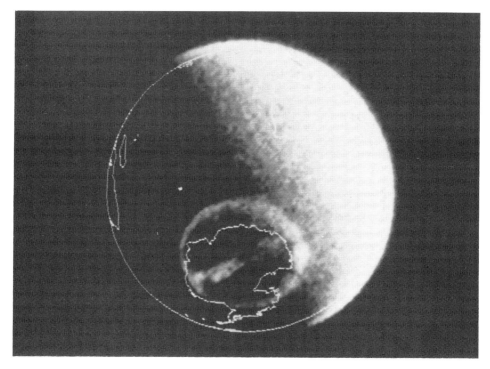

Fig. 6.8. A Dynamics Explorer 1 view of the southern auroral oval from high above Antarctica, showing the oval in 'theta' configuration. Theta aurora comprises a bridge of relatively weak activity running along the noon–midnight meridian *within* the oval, and is occurs at times of low geomagnetic activity. Courtesy of University of Iowa.

the evening sector theta aurora have been made from Spitsbergen (Simmons and Henriksen, 1992).

Theta aurora is produced by low energy (0.6–1.0 keV) electrons from the magnetotail. These electrons do not undergo acceleration in near-Earth space, and are therefore less penetrative than those giving rise to classical polar aurora; the base altitude of theta aurora is typically around 150 km.

Plasma movements in the transpolar arc system, when it is present, appear to reflect circulation in the lobes of the magnetotail (section 5.2), with an anti-sunward flow on the evening side, and sunward flow at the dawn side. In the evening sector, the theta aurora 'bridge' divides into east and west branches where it meets the auroral oval around the Harang discontinuity (section 5.3).

Visually, theta aurora appears faint and relatively quiescent, comprising long, thin, more or less static rays. Ray tops may reach into the sunlit upper atmosphere, where blue-violet N_2^+ emissions at 391.4 nm and 427.8 nm are seen. Predominantly, however, theta aurora emission comes from the auroral green OI 557.7 nm line. These discrete arc structures produced in quiet times contrast with the diffuse polar glow aurora present within the polar cap at times of high geomagnetic activity (section 6.4.2).

REFERENCES

Davis, N. (1992) *The aurora watcher's handbook*. University of Alaska Press.

Elvey, C. T. (1965) Morphology of Auroral Displays. In: Walt, M. (Ed.), *Auroral phenomena*. Stanford University Press.

Hey, T., and Walters, P. (1987) *The quantum Universe*. Cambridge University Press.

Hultqvist, B. (1969) Auroral and polar cap absorption. In: McCormac, B. M., and Omholt, A., (Eds), *Atmospheric emissions*. Van Nostrand Reinhold.

Lassen, K. (1969) Polar cap emissions. In: McCormac, B. M., and Omholt, A., (Eds), *Atmospheric emissions*. Van Nostrand Reinhold.

Livesey, R. J. (1989) The aurora. *Meteorological Magazine* **118** 253–260.

Mitra, S. K. (1952) *The upper atmosphere* (2nd edn). The Asiatic Society, Monograph Series Vol. V.

Moore, P. (1979) *The Guinness book of astronomy*. Guinness Superlatives.

Ratcliffe, J. A. (1970) *Sun, Earth and radio: an introduction to the ionosphere and magnetosphere*. Weidenfeld & Nicolson.

Sandford, B. P. (1967) High latitude night-sky emissions. In: McCormac, B. M. (Ed.), *Aurora and airglow*. Reinhold.

Simmons, D. A. R. (1985) A study of auroral emissions by interference filter photography. *J. Brit. Astron. Assoc.* **95** 252–256.

Simmons, D. A. R. (1988) Auroral photography at high latitudes. *J. Brit. Astron. Assoc.* **98** 93–97.

Simmons, D. A. R., and Henriksen, K. (1992) Discrete polar cap aurora observed from Spitsbergen. *Polar Record* **28** (166) 191–204.

Simmons, D. A. R., and Henriksen, K. (1994) Daytime aurora observed from Spitsbergen. *Polar Record* **30** (173) 85–96.

Simmons, D. A. R., and Henriksen, K. (1995) Polar-glow aurora observed from
    Spitsbergen. *Polar Record* **31** (178) 315–326.
Vallance Jones, A. (1969) Auroral spectroscopy. In: McCormac, B. M., and Omholt, A.,
    (Eds), *Atmospheric Emissions*. Van Nostrand Reinhold.
Vallance Jones A. (1991) Overview of auroral spectroscopy. In: Meng, C.-I., Rycroft, M.
    J., and Frank, L. A. (Eds) *Auroral physics*. Cambridge University Press.

7

The mid-latitude aurora

7.1 CAUSES OF MID-LATITUDE AURORAE

While traditionally regarded in the popular astronomical literature as a phenomenon strictly of high latitudes, the aurora can, under certain circumstances (section 5.4.2.2), extend to lower latitudes and become visible to observers in the more populous regions of the world. Occasional great displays, as in 1938 and 1989, may be sufficiently intense and widely seen to elicit considerable media and public interest (section 2.6).

Such events not only provide a spectacular light display for the observer: their associated electrical ground currents may also trigger surges in power grid systems, causing disruption of supplies as experienced in areas around Quebec in March 1989, or—less dramatically—present problems for owners of automated garage door systems, as in California in August of the same year. Computer malfunctions connected with the same activity brought the Toronto stock market to a halt (Dayton, 1989). These ground-level effects can provide a key to early detection of possible visible auroral events.

Ground-based observations can reveal the overall pattern of activity within an auroral storm, although only a relatively small section of the auroral oval in the observer's hemisphere may be viewed from a single location. As discussed previously, imaging of the auroral ovals in their entirety had to await the ISIS and Dynamics Explorer missions of the 1970s and 1980s. It is a tribute to the ground-based observers of the IGY and earlier in the twentieth century that the prediction of the auroral ovals' existence based on their work was confirmed in the advent of the Space Age.

The frequency with which the aurora may be seen from a given location is influenced, principally, by geomagnetic, rather than geographical, latitude. Consequently, observers in the United States see aurorae more frequently than their colleagues at equivalent geographical latitudes in northwest Europe: the present location of the north geomagnetic pole near Greenland favours American observers. For a given geographical latitude, an observer in the United States lies at a higher geomagnetic latitude than one in Europe. For example, one of the world's most productive aurora observers (having logged no fewer than 1000 nights of activity up to the summer of 1992!) is Jay Brausch, of Glen Ullin in North Dakota. Glen Ullin has a geographical latitude of 46°48′, but a geomagnetic latitude of 56°. Brausch makes frequent reports of aurorae which are invisible to northwest European observers during the summer twilight. Weather conditions are also apparently more favourable in North Dakota. In recognition of his work, Jay Brausch was awarded the

Merlin Medal of the British Astronomical Association in 1993; a gallery of his photographs appears in the BAA Aurora Section's 1993 report (Livesey, 1994).

A study of more then 30 years' auroral reports from visual observers in the United Kingdom by Ron Livesey, Director of the BAA Aurora Section, suggests an average frequency of displays of 5 per annum in the south of England, reaching 60–70 per annum in the meteorologically favoured location of the Moray Firth in northeast Scotland (Livesey, 1995). Other factors, such as weather and summer twilight, have to be taken into consideration, naturally. Many naive amateur astronomers from the south of the UK have travelled in vain to view the aurora from Scotland at the height of midsummer, and Scots observers will certainly aver that, contrary to popular imagination, the aurora is, sadly, not a nightly feature of their skies!

Fig. 7.1 indicates the orientation of lines of geomagnetic relative to geographical latitude. Rounded geomagnetic latitudes for some geographical points are listed in Table 7.1. Southern hemisphere observers are at something of a disadvantage, since the bulk of the southern land-masses lie far from the regions of maximal auroral occurrence.

Table 7.1. Some geographical locations and their geomagnetic latitudes

Location	Geographical latitude (°N)	Geomagnetic latitude (°N)
Los Angeles	34	45
New York	41	52
Montreal	46	56
Anchorage	61	61
London	51	54
Edinburgh	56	58
Aberdeen	57	59
Reykjavik	64	69
Copenhagen	55	55
Paris	49	51
Madrid	40	44
Rome	42	40
Moscow	55	53

Observers in the northeast of America are, by virtue of their proximity to the north geomagnetic pole, at something of an advantage over those at similar geographical latitudes in Europe.

The auroral storms which penetrate to middle latitudes (those of the central United States, Australasia, and northern Europe including the British Isles) arise from disturbances of the normal, quiet auroral conditions which prevail for much of the time. These disturbances mark the arrival in near-Earth space of energetic particles injected into the solar wind following flares associated with active sunspot groups, or the Earth's passage through a steady, higher-intensity particle stream ejecting into the solar wind from a coro-

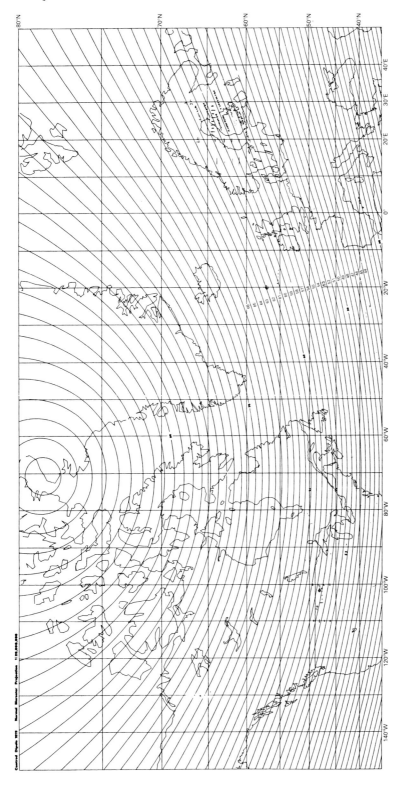

Fig. 7.1. Alignment of geomagnetic versus geographic latitude in the north Atlantic region. Note particularly how, as a result of their proximity to the north geomagnetic pole, observers at relatively low geographic latitudes in the northeastern United States are at higher geomagnetic latitudes than their European colleagues.

nal hole. In either case, it is the direction of the prevailing Interplanetary Magnetic Field which determines whether the aurora will be brought to lower latitudes.

When the IMF in the Earth's neighbourhood has a strong southerly component relative to the ecliptic, energetic particles from the magnetotail can be injected into the high atmosphere as a result of reconnection between terrestrial and solar wind field lines. Models suggest that the injection of particles into the high atmosphere via this route may either be continual or spasmodic. In the first instance, moment-to-moment changes in auroral structure could be taken as a direct reflection of the continually changing local IMF. An alternative view is that spasmodically entering solar wind particles can accumulate in the magnetotail, generating stress on the magnetosphere which is eventually released by an earthwards surge of particles, in turn giving rise to enhanced auroral activity. The latter mechanism may work on a 40–60-minute cycle, resulting in regular outbursts during an observed storm.

A major consequence of the arrival of accelerated solar particles from the magnetotail is the broadening and expansion of the evening sector of the auroral ovals, accompanied by surges of activity running westwards around the ovals and an increase in the energy of the auroral dynamo. Particles trapped in the Van Allen belts also begin to circulate more rapidly, contributing to the overall weakening of the global magnetic field. This disturbance allows the aurora to temporarily reach lower latitudes.

7.2 APPEARANCE AND BEHAVIOUR

The sequence of forms and activity witnessed by an observer on the ground during an auroral event penetrating to lower latitudes will, generally, reflect the global behaviour of the auroral ovals. A display's brightness, activity, and extent in the observer's sky will depend largely on two factors. The first of these, the geomagnetic latitude of the observing site, has already been discussed (section 7.1). The *cause* of a particular auroral event will also determine its extent and activity: aurorae resulting from the arrival of turbulent, highly magnetized pockets of energetic plasma in the solar wind following a flare event or coronal mass ejection will be far more active and extensive than those brought about by Earth's immersion in the more quiescent particle streams injected into the solar wind via coronal holes in the declining years of the sunspot cycle.

The French solar physicists J. P. Legrand and P. A. Simon have carried out studies of amateur and other auroral observations collected since the International Geophysical Year (Legrand and Simon, 1988). Their research suggests that the main source of auroral activity at lower temperate latitudes is from solar flares during the early parts of the sunspot cycle. On this basis, historical records of auroral activity may indeed be a reliable means of determining solar maxima in pre-telescopic times, as suggested by Eddy and others (section 2.3).

Fig. 7.2 presents a plot of the occurrence of very major geomagnetic storms—defined, in this instance, as readily visible (or at least potentially so in clear skies) to the latitude of the Channel coast of England—over solar cycles 21 and 22. In the main, these are clustered around the respective sunspot maxima, in 1980 and 1990. The existence of a double peak in auroral activity is evident to some degree: most major storms occur about a year ahead of sunspot maximum, and in an interval some 12–18 months after maximum. In

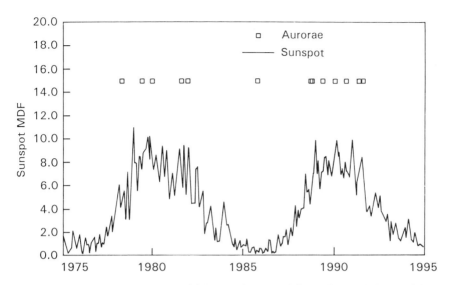

Fig. 7.2. Major geomagnetic storms visible (or at least potentially so) from the latitudes of the English Channel coast during sunspot cycles 21 and 22. These major events cluster around the sunspot peaks, with the notable exception of the February 8–9 1986 display. Isolated occurrences such as this place some constraints on the viability of using historical records of auroral activity as indicators of the overall level of sunspot activity in pre-telescopic times.

cycle 22 the first peak in auroral activity penetrating to low latitudes occurred in 1988–89, with a secondary peak in 1991 (section 7.2.4).

Fig. 7.2 also highlights one of the perils of using historical auroral records as an index of past solar activity. Of the 14 major storms in the 21-year period period from 1975 to 1995, 13 can be clearly associated with the time of highest sunspot activity. Historical auroral records might therefore allow the timing of pre-telescopic sunspot maxima to be estimated to within an accuracy of two years or so. The obvious problem, however, is the occurrence of events such as that of 1986 February 8–9 (section 2.6), appearing in splendid isolation almost on the sunspot *minimum* between cycles 21 and 22. One-off events such as this of necessity reduce the likely accuracy of any determination of past solar activity based upon auroral records. Clustered, multiple sightings in a given year *might* be taken as more reliable indicators of maximum activity.

The great mid-latitude displays of 1938, 1957, and 1989 all had their origins in violent solar activity associated with the rise towards sunspot maximum. Such displays are characterized by rapid changes in form, and brightness, and produce extensive rayed structures, sometimes across the entire sky even from such low-latitude locations as southern England or the southern United States. The coronal-hole-induced aurorae, however, tend to consist of quiet diffuse glows, or homogeneous arcs and bands, with only occasional outbursts of rays. Particularly at those locations in the temperate regions where the observer enjoys a reasonably high geomagnetic latitude (as in northern Scotland, or the northeastern United States), both forms of activity can be seen at the appropriate times in the sunspot cycle. While no two aurorae are ever identical, a broad outline of the patterns of activity to be expected by the observer on the ground on a typical active night, depending on the solar activity which has triggered the aurora, can be given (section 7.2.1; section 7.2.2).

7.2.1 Solar-flare-induced aurorae

The most spectacular aurorae at mid-latitudes are certainly those which follow the arrival, in near-Earth space, of energetic plasma ejected into the solar wind during a solar flare. Flares are usually associated with large, actively changing sunspot groups, which are often commonest in the early parts of the cycle. As discussed later (section 7.2.4), predicting the occurrence of aurorae at mid-latitudes is very difficult, but it might generally be said that the chances of witnessing an active display from a relatively low latitude site are best during the year or so ahead of sunspot maximum itself.

A visual auroral display resulting from a solar flare event will often begin in early evening as a quiet *glow*, low over the polewards horizon. Clouds can often appear in silhouette against the glow, while stars can be seen quite clearly through it. Where the geomagnetic disturbance is relatively small, producing only a slight expansion of the auroral ovals, such a glow might represent the maximum extent of the auroral display: obviously, observers at higher geomagnetic latitudes will enjoy a more extensive display under such circumstances, while the lower latitude observer merely witnesses the uppermost parts. Glows are often faint, sometimes little more obvious than the Milky Way, and will often be missed, especially in areas badly afflicted by artificial light pollution, or in bright moonlight.

Observers must be cautious to avoid misidentifying glows produced by artificial lights in towns or villages along the line of sight to the pole. Especially on slightly hazy nights, spurious auroral reports due to light pollution are not uncommon.

Auroral glows may fade away after a time, without anything further being seen. On occasions when the aurora is likely to become more active, however, the glow may brighten and rise higher into the sky. From this stage, a more definite structure may develop in the aurora. The glow often resolves itself into an *arc* structure spanning the poleward sky. The arc, which has been likened by some observers to a white or pale green rainbow, will usually show no internal structure in the early part of the display, and in such a condition is described as *homogeneous*.

The highest point on the base of an arc lies roughly in the direction of the observer's magnetic pole. Thus, for observers in the British Isles, for example, the highest point on auroral arcs is usually a little to the west of north. Reflecting the sharp lower boundary for auroral emissions in the high atmosphere, the base of the arc is much more sharply defined than the often rather diffuse upper extremities.

After a time, a homogeneous arc will often develop bright areas along its length, from which vertical *ray* structures appear. *Fading* of the arc may also be the prelude to ray formation. The rays can be fairly static, though in an active display, some degree of movement is more normal. On occasion, the observer may see only isolated bundles of rays, whose appearance can be likened to searchlight beams. In a brightening display, colour may become more evident, and the differing red and green oxygen emissions along the vertical extent of the rays can sometimes be very striking: green emissions predominate low down, red higher up (section 6.2). By this stage, the aurora may be quite obvious, even from light-polluted locations.

Folding of the arc upon itself produces a ribbon-like *band*, which again may be homogeneous or rayed. Rapid movement of rays along the length of a band produces the 'curtain' illusion, commonly illustrated in popular astronomy books. The illusion is particularly effective in events where the rays are long, and fill much of the poleward sky.

Rarely, at the very climax of an extremely violent auroral display, activity may pass overhead and into the equatorwards half of the observer's sky. At this stage, the rays and other features will appear to converge on a single area of the sky as a result of perspective, producing the form of a *corona*. The corona is normally centred, at mid-latitudes, some degrees equatorwards of the true zenith in the observer's sky, near the *magnetic zenith*.

In a typical storm, the corona—if it forms at all—is usually short-lived, particularly at lower latitudes. Auroral activity will normally fall back towards the polewards horizon within a few minutes. There, the aurora may remain as a glow for a time before fading away.

Sometimes, especially during a particularly violent storm, the aurora may continue to show rayed activity, rising to coronal peaks several times in the course of the night. The intense aurora of 1989 March 13–14 was remarkable in showing coronal forms more or less throughout the night at higher temperate latitudes. It is also possible for an aurora which has fallen back to a glow to go through the entire sequence from glow through rayed to coronal forms several times on an active night (Gavine, 1979).

Other forms of aurora can accompany the arcs, bands and rays. The sky may be suffused by a weak background *veil* during some displays, and *patches* (or surfaces) of homogeneous auroral light outlying the more structured parts of the display can also be seen. Unusual aurorae comprising only diffuse patches have been reported, an example being the display witnessed by British observers on the night of 1978 August 29–30: such events might frequently be the source of UFO reports.

Mid-latitude aurorae are subject to variations which can be of long or short timescales. Arcs, bands and patches often exhibit slow *pulsations* over the course of several minutes. *Flaming* activity, in which waves of light sweep rapidly upwards from the horizon, locally brightening features as they pass, is much more rapid, with several waves per second at its most intense. Though most often seen in the declining phase of an aurora, flaming sometimes precedes corona formation. *Flashes* resembling lightning are sometimes reported, while intense aurorae may show *flickering* as light passes horizontally through features.

Movement also takes place over differing timescales. Quiescent features may shift their position in the sky only very slowly and subtly. Rays are often seen to drift along the length of arcs and bands. A particularly eerie effect of the aurora is the slow, silent drift of isolated ray bundles above hills or other features on the polewards horizon, as viewed from a remote, dark location. Folding and rippling of a rayed band strengthens the illusion of 'curtain' structure within an auroral display. Rapid movement of rays along an arc or band produces the effect of streaming. Coronal rays can show a 'cartwheel' rotation around the centre from which they appear to radiate.

The vigorous aurorae at mid-latitudes which may follow solar flare events tend to last only for one or two nights at most. There will usually be a single night of particularly intense activity, perhaps flanked by nights of less major aurora. Unlike the coronal hole aurorae (section 7.2.2), flare events tend to be seen only at a single solar rotation; recurrences are quite rare, since the active sunspot groups above which flares develop have, themselves, fairly short lifetimes.

7.2.2 Coronal hole aurorae

In contrast with the often violent, active events of the rising years of the sunspot cycle, those few aurorae seen from mid-latitudes late in the cycle tend to be relatively quiet. These late-cycle events are normally produced by the passage of broad, persistent coronal hole particle streams, which open up particularly a couple of years prior to sunspot minimum (section 4.3.5).

Coronal holes may persist for months on end, and an individual particle stream may sweep over the Earth several times, resulting in auroral displays which recur at 27-day intervals. Such recurrences might be more reliably forecast than the capricious flare ejections from sunspot groups. For example, a persistent coronal hole was the source of a long series of recurrent mid-latitude aurorae throughout the second half of 1985.

As with the flare-induced aurorae, those events produced by coronal hole streams may often commence as a featureless glow on the poleward horizon. Activity may rise higher into the sky, assuming the form of arcs or bands, with occasional rayed outbursts. Movement and changes of brightness tend to be less marked in coronal hole aurorae, and it is not unusual for such activity to progress no further than the glow or quiet arc stage.

The expansion of the auroral ovals resulting from passage of a coronal hole stream is typically smaller than that which follows arrival of solar flare-derived particles. The relatively quiescent coronal hole aurorae are therefore usually best seen from higher latitudes. Coronal hole aurorae probably figure far less frequently than flare aurorae in the historical records from lower latitudes (section 7.2.1).

7.2.3 'Flash aurora'

Among reports collected by the BAA Aurora Section are a number of instances where observers record isolated, short-lived bursts of auroral activity. The duration of these can be very brief indeed—for example, one event, reported by an experienced observer (Todd Lohvinenko, Winnipeg, Canada) in 1987 lasted only five seconds, but passed through the zenith (Livesey, 1988). Similar reports, albeit of less extensive activity, have been made by the vastly experienced Scottish observers Alastair Simmons and Dave Gavine, lending support to the authenticity of these occurrences. Magnetometer records may also lend support: some of these 'flash aurora' events can be correlated with minor magnetic field fluctuations.

While further observations are required to absolutely verify the occurrence of these transient events, professional geophysicists have already offered explanations for them. One suggestion is that flash aurorae are correlated with high-latitude activity, and follow substorms (section 5.4.2.1) by about four hours. Reconnection processes in the magnetotail have been suggested as the causative mechanism.

7.2.4 Auroral prediction

The aurora often takes the mid-latitude observer by surprise, thanks partly to its reputation as a strictly polar phenomenon, and partly because it is a phenomenon whose occurrence

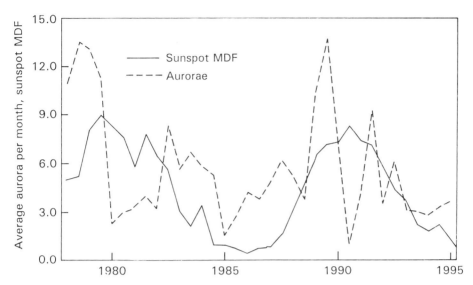

Fig. 7.3. Comparison between the averaged monthly frequency, over 6-month intervals, of aurorae observed from the British Isles south of the Orkneys, and 6-monthly means of sunspot activity over the same periods, in the years between 1978 and 1995. Note the peaks of auroral activity ahead of sunspot numbers in cycle 21, which came to maximum in 1979, and in cycle 22 which peaked in 1990. Secondary auroral peaks—which may in part be ascribed to coronal hole activity—are seen in the declining years of either sunspot cycle.

is extremely difficult to forecast. Early warning of the likely occurrence of aurorae at lower latitudes may be obtained by careful study of changing sunspot activity, or through the operation of radio or magnetometer equipment.

Long-term studies have shown that the numbers of active aurorae penetrating to lower latitudes are greatest during the rapid climb in sunspot numbers a year or so before sunspot maximum (Livesey, 1985; Fig. 7.3). Using even simple equipment to safely observe the Sun by the method of projection, amateur astronomers are able to monitor day-to-day changes in white-light sunspot activity, while professional and advanced amateur observers can follow the Sun's behaviour at other wavelengths of the electromagnetic spectrum, notably that of hydrogen-alpha emission.

Radio noise, and bursts associated with solar active regions, (section 4.3.6) may be followed using appropriate equipment (Ham, 1991).

Plots of the month-to-month variations in sunspot numbers and the frequency of aurorae at low latitudes show considerable variation, making the broad activity pattern difficult to ascertain; a somewhat clearer picture is obtained by taking 6-month averages for both numbers, as has been done to produce Fig. 7.3. From the figure, it becomes apparent that low-latitude aurorae reach their peak frequency of occurrence about a year ahead of sunspot maximum, and have a secondary peak following sunspot maximum: the pattern is broadly similar to that for the very major global auroral storms shown in Fig. 7.2.

The presence of large, actively changing sunspot groups near the central meridian of the projected solar disk should alert the observer to the *possibility* of auroral activity penetrating to lower latitudes in the following 24–36 hours. Solar flare activity (which may be directly observed in hydrogen-alpha light) is frequently associated with such groups, but there is no guarantee that aurora will necessarily follow. For example, high sunspot numbers in 1980, near the maximum of solar cycle 21, yielded scarcely any aurorae for observers at the latitude of central Scotland.

In many respects, predicting the widespread, readily visible occurrence of the aurora is on the same level of difficulty as forecasting the brightness of comets—amply illustrated by such public disappointments as Kohoutek in 1973 or Austin in 1990. An example of the aurora failing to live up to expectations followed extensive alerts in the British media in June 1989 after a vigorous flare had been observed in a complex naked-eye sunspot group. In the event, no aurora was seen, to the disappointment of the public, many of whom had missed the Great Aurora of the previous March: curiously, the same media which publicized the June possibility (which could hardly have met less favourable circumstances at the height of summer twilight, full Moon, and cloudy skies!) had been reluctant, on the night, to alert their audience to the occurrence of the March event. A similar alert, issued in June of 1991, proved equally fruitless for observers in the United Kingdom, though a fairly short-lived outburst of high activity was seen from the United States (section 7.8)

Recent records of auroral sightings can be of some value in attempts to anticipate aurorae in the immediate future. Where available, such reports may be plotted onto a Bartels chart, named after the early twentieth-century German solar physicist and student of the aurora: Fig. 7.4 presents a simple example. Time is shown in vertical 27-day strips, each roughly corresponding to a single rotation of the Sun about its axis as viewed from Earth. Auroral events, coded for intensity, are plotted at the appropriate date on the chart, where recurrences will be revealed as they line up in adjacent strips. Particularly in the later parts of the sunspot cycle, as coronal holes become common, series of aurora spanning several rotations may be seen. Flare-induced events show less tendency to recur, as the spot groups from which they arise are relatively short-lived: sunspot-associated aurora tends also to be less active and extensive in the event of any recurrence. Ideally, an observer with awareness of recent activity should be particularly watchful for possible recurrences 27 days after major events at lower latitudes.

More complicated versions of the Bartels chart can incorporate details of radio aurorae or geomagnetic field disturbances. The simple version in Fig. 7.4 does have the advantage of appearing uncluttered.

Amateur radio operators have long used the clouds of enhanced high-atmosphere ionization generated by auroral activity as reflectors from which to propagate short-wave signals over greater-than-normal distances. Forecasts of times when favourable conditions for long-distance communication are likely to prevail are obviously of value to this community. Weekly updated forecasts of propagation conditions are broadcast for UK radio operators. As in visual forecasting, much emphasis is usually placed on the pattern of activity in the previous 27 days and the likelihood of its repetition. In the United States, the WWV radio station's short-wave broadcasts carry a geomagnetic forecast at 18 minutes past each hour (Sampson, 1992).

Also in the United States, the NOAA produces a weekly bulletin, *Preliminary report and forecast of solar-geophysical activity*, published from the solar observatory at Boulder, Colorado. Again, recent observations form the basis for the bulletin's geomagnetic forecasts. Publication of such forecasts by a government agency is recognition of the relevance of auroral effects to both navigation and communication. 'Interplanetary weather'

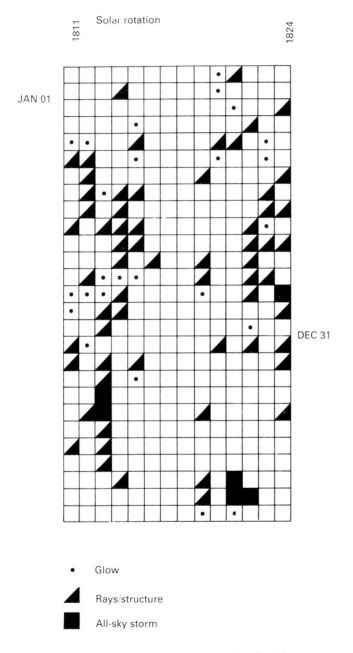

Fig. 7.4. Comprehensive Bartels diagram for 1989, summarizing observations reported to the Aurora Sections of the British Astronomical Association and Junior Astronomical Society. Time is plotted in 27-day strips, each roughly corresponding to a single solar rotation. Recurrent events, potentially resulting from continued activity in disturbed solar regions, appear to line up in adjacent strips.

forecasts are also issued via the World Wide Web, accessible through suitably connected personal computers.

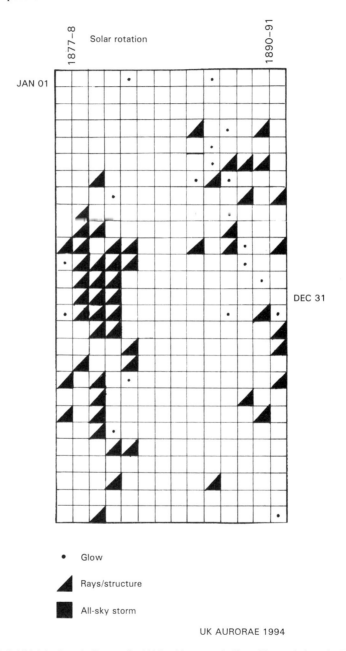

Fig. 7.5. British Isles Bartels diagram for 1994, with aurorae indicated by symbols as for Fig. 7.4. By this time, late in cycle 22, major geomagnetic storms were rare, and activity was dominated by recurrent coronal-hole type events at the latitudes of north Scotland. Particularly in the opening months of the year, recurrent periods of activity (and, equally, low activity) are clearly seen.

Early warning of the actual occurrence of auroral activity can be obtained through detection of associated ionospheric or geomagnetic effects. The detection, typically in mid-afternoon, of enhanced short-wave radio propagation conditions can indicate the onset of auroral activity. It is a problem, however, that not all radio aurorae have a visual counterpart, and vice versa.

Disturbed geomagnetic conditions may be directly measured using magnetometers: even those lacking electronics skills should be able to construct the very simple 'jam-jar' form, which has apparently enjoyed widespread use (section 8.2.1). Magnetic disturbances may provide an early warning of the aurora's possible visibility, but need not necessarily coincide with events visible from the observer's latitude. Geomagnetic disturbances are often detectable at lower latitudes than any accompanying visual display. If very major magnetic fluctuations are observed, however, it is likely that visual observations should follow if skies are clear.

In the wake of the 1989 March storm, many local astronomical societies and clubs in the United Kingdom set up 'ring-round' telephone networks, which come into operation when major auroral activity is seen. Similar networks exist in the United States, notably the Aurora Alert Hotline. Such early-warning systems can work well, provided no single participant ends up making telephone calls during time more effectively spent making observations!

Perhaps the safest conclusion to be drawn regarding the forecasting of aurora at mid-latitudes is that the visual observer hoping to see a display will need to show patience and perseverance. The majority of reported events are probably detected by astronomers engaged in other forms of work. Amateur meteor or variable star observers, for example, may suddenly become aware of activity as it increases during a major display in the course of an evening. Systematic checks on the polewards part of the sky from a dark location on every possible clear night will eventually be rewarded, and seem preferable to any reliance on forecasts.

7.2.5 Visual observations
The advent of spacecraft monitoring of the auroral ovals from above has undoubtedly improved the precision with which the global pattern of auroral storms can be followed. Some value is still placed on visual reports, however, and these are collected by a number of organizations, such as the Aurora Sections of the British Astronomical Association and Royal Astronomical Society of New Zealand. Modern visual observations may be compared with those in the archives, for example, allowing their correlation with data from pre-satellite times. Such work can reveal whether auroral behaviour has changed significantly in historical times.

Annual summaries of auroral observations are compiled and published by the BAA Aurora Section (for example, Livesey, 1989, 1990). Reports are also published in the *Marine Observer*, produced by HMSO in the United Kingdom.

It is to the advantage of the observer that the aurora shows a fairly small range of typical forms (Paton, 1973; Gavine, 1995; section 7.2.1). A standard reporting code has been in use for many years, allowing the state of auroral activity at any given moment to be accurately and concisely noted in a form useful to those wishing to derive positional and other information. The *International Auroral Atlas* provides photographic examples of the aurora's typical appearance. The examples provided in the plates here show the series of structures which might be expected during the development of auroral activity at mid-latitudes (sections 7.2.1, 7.2.2).

Visual observers submit reports, accurate to the nearest minute or so, of the type(s) of aurora present on a given night (always quoted as a double date, for example 1990 April 13–14, to avoid ambiguity), and the extent of these types in altitude and azimuth. Azimuthal extent of auroral forms is estimated in degrees, with 0 due north, 090 due east, and so on. The altitude in the observer's sky of the highest point on the base of a feature such as an arc or band, denoted by the symbol h, is particularly useful. Since aurorae have a fairly sharply defined lower boundary 100 km above the Earth's surface, h and the observer's known latitude and longitude may be combined to provide two elements of a pythagorean triangle, and the latitudinal extent of an auroral feature can therefore be easily deduced. Combination of several such reports, obtained simultaneously, affords greater accuracy in this method of triangulation.

A second altitude measurement, denoted by the symbol ↗, denotes the upper edge of an auroral feature. This is sometimes difficult to determine, as the tops of rays and other features are often diffuse and difficult to discern. Fig. 7.6 summarizes the positional measurements which may usefully be recorded by the visual observer.

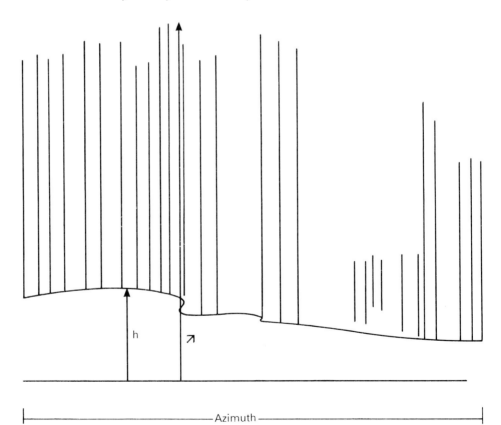

Fig. 7.6. Schematic representation of the useful altitude and azimuth measurements which may be recorded by a visual observer during an auroral display.

Table 7.2. The auroral brightness scale

(i)	Weak, comparable in intensity to the milky way.
(ii)	Comparable to moonlit cirrus clouds.
(iii)	Quite strong aurora, comparable in brightness to moonlit cumulus clouds.
(iv)	Stronger than (iii), possibly even bright enough to cast shadows. Aurorae this bright are relatively rare at mid-latitudes, but have been reported during great storms as in March 1989.

It is usually sufficient to make note of the condition of an auroral display only as and when it changes. There is little point in making estimates of altitude and azimuth of a quiescent arc at intervals more frequent than about 15 minutes. Rapidly changing aurorae, such as those which follow many solar flares, present a challenge to the observer, and will

Table 7.3. Standard code for auroral recording

Condition: Aurorae may be either quiet (q), with no movement, or active (a). Activity may take several forms, denoted by subscripts.
a_1 Folding of bands.
a_2 Rapid change of shape in lower structure.
a_3 Rapid horizontal movement of rays.
a_4 Fading of forms, with rapid replacement by others.

Changes in brightness are also seen, described by p followed by a subscript.
p_1 Slow pulsing.
p_2 Flaming; waves of light passing vertically through display.
p_3 Flickering; rapid, irregular variations.
p_4 Streaming; irregular horizontal variations in homogeneous forms.

Form: The aurora shows a range of discrete structures. These may be homogeneous (H), lacking internal structure, or rayed (R), showing vertical structure of varying lengths. Subscripts denote the lengths of rays from R_1 (short) to R_3 (long). Features may be multiple (m), fragmentary (f), or coronal (c).

G Glow with no other structure, often lying low above poleward horizon.
A Arc structure, an arch of light spanning east–west across the sky. May be homogeneous (HA) or rayed (RA).
B Folded, ribbon-like structure, often developing from arc. May have HB or RB
R Rays may sometimes be seen in isolation (eg R_1R), or in bundles (eg mR_2R) when no other aurora is present.
V A background veil which sometimes pervades the sky during auroral displays.
P Patches. Discrete areas (sometimes referred to as 'surfaces') of auroral light. May appear as HP or RP.
N Auroral light, not identifiable, seen for example through cloud.

Colour: Characteristic colours exhibited by the aurora may be described by use of lower-case letters at the end of feature descriptions.
a Red in upper parts of aurora only.
b Red in lower border of aurora only.
c Green, white or yellow.
d Red.
e Red and green together.
f Blue or purple.

Examples: qHA1c describes a quiet (q) homogeneous arc (HA) of brightness 1, white in colour.
$a_3p_2mR_2B3e$ describes a bright red and green multiple rayed band, with medium-length rays showing horizontal movement and flaming activity.

in all probability change too quickly for anything but the broad pattern of activity to be noted.

Changes in brightness can be recorded, and may reflect impending changes in activity. The onset of flaming, for example, may precede corona formation in a major storm, while the brightening or fading of an arc may indicate the imminence of increasing activity. A four-point scale is used to denote brightness (Table 7.2).

The accepted standard code used to concisely describe auroral forms and their state of activity is listed in Table 7.3, while Table 7.4 gives an actual example of the code's usage from the Author's observing log.

A minimum of equipment is required to make useful visual observations of the aurora. The naked eye suffices, and optical equipment such as binoculars will serve merely to diffuse the structure of any aurora present. The accuracy of visual altitude estimates may be improved by use of an alidade which, in its simplest form, consists of a protractor attached to a plumb line and some sort of sighting device (Fig. 7.7). When the auroral feature to be measured is observed through the sighting device, a direct measure of alti tude relative to the horizon is obtained by reading the angle of the plumb line against the protractor's scale.

Table 7.4. Sample auroral observation.

Observer: N.M.Bone. Location: Campbeltown, Scotland.

Lat. 55°25′N Long. 5°36′W

April 10–11, 1982.

		h	↗	Azi	Remarks
2126 UT	qHA2c	15	25	270–060	Dark sky below.
	V1		30–35		
2132	p$_1$HB	20			
2135	mR$_1$R1–2c		20	320–330	Short-lived.
2147	aR$_1$A1–3	5		320–005	Very sharp rays.
	R$_2$R				To west of arc.
2155	R$_2$B				
2159	a$_3$mR$_3$B3e	22	90		Rayed forms across entire northern sky, starting to converge at tops.
2213	aR$_3$B2–3	10	60	260–000	Very dark below.
2222	R$_3$R			260–355	Becoming quieter, bulk of activity in west.
2232	aR$_3$B3e	10	55	280–060	
2235	aR$_2$A	15	25	280–060	Quieter.
2243	R$_2$A1–2	8	35	280–060	Much quieter, Moon now becoming a problem.
2340	G2		20	000	Light still present. End of observations.

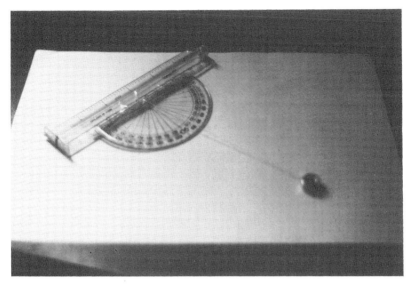

Fig. 7.7. A simple alidade, basically consisting of a plumb line and protractor scale aligned with a sighting device, allows reasonably accurate determination of elevations of auroral features, and other objects in the sky.

7.2.6 Photographic observation

Visual observations of the aurora can reveal much about patterns of activity and approximate positional information during displays. More accurate, permanent positional records can readily be obtained through photography, for which even small cameras are adequate. Only simple equipment, widely available in the amateur astronomical community, is required. The results, especially on colour films, can often be extremely aesthetically pleasing.

The 35-mm SLR cameras in widespread use are eminently suitable for auroral photography, provided these have the facility for taking time exposures in excess of a couple of seconds via either a 'B' or 'T' setting, preferably operated by a cable release mechanism. The camera should also, ideally, be firmly mounted on a tripod. To be useful for triangulation work, auroral photographs should be timed accurately, to within a couple of seconds, as demanded by the frequently rapid alteration in auroral forms.

There are few hard-and-fast rules for auroral photography. Most observers prefer to use fast colour slide films, typically of ISO 400. Kodak and Fuji emulsions have been used with considerable success (for example, Simmons, 1985), and seem less prone to the garish colour enhancements produced by other films. It is certainly a common characteristic of photographic emulsions that they will bring out auroral colours more marked than those visible to the eye, and greens and reds are often particularly enhanced. Most observers prefer to use the camera at full aperture, preferably $f/2$ or faster. Exposure time depends on the brightness of the auroral forms to be photographed.

Using ISO 400 film, weak horizon glows will be photographed with exposures of 30–60 seconds at $f/2$. Moderately bright auroral structures will record in 20–30 seconds at *f*/2, while the shutter need only be opened for 5–10 seconds at *f*/2 during bright, actively moving aurorae. Indeed, in the latter case, detailed structure will be lost due to auroral movement if the shutter is left open too long.

Perhaps the best advice which can be offered to those wishing to photograph the aurora is to take plenty of bracketed exposures. Particularly at lower latitudes, it may be some time before a further opportunity to photograph the aurora arises, and the author would therefore suggest that observers in such locations as the south of England or the southern United States do not skimp on film when a good display does occur!

The sensitivity of photographic emulsions is often such that sub-visual aurorae can be recorded. During a quiet interlude in the November 1991 storm, for example, the author succeeded in recording diffuse red auroral emission over much of the northern sky from Sussex in the south of England using a 2-minute exposure at *f*/2.8 on ISO 400 film, although no aurora was evident to the naked eye. Similar results were inadvertently obtained by deep-sky photographer Ron Arbour from Hampshire (also in southern England) during the March 1989 storm; exposures taken for galaxies in a spell when the aurora seemed absent were fogged by the sub-visual aurora suffusing the whole sky at this time.

Considerable scope obviously exists for auroral triangulation work using simultaneous photography by widely stationed observers. Dr Michael Gadsden of Aberdeen University has suggested that observers should take photographs of any aurora visible from their location, irrespective of auroral form or condition, *precisely* on the hour, half hour and intermediate quarter hours. If all observers were to adhere to this method, the number of simultaneous auroral photographs suitable for triangulation work would be increased greatly.

7.3 CONTRASTS WITH HIGH-LATITUDE AURORAE

The aurorae penetrating to lower latitudes have their origins in disturbances of the normal pattern of global magnetism, and a resulting equatorwards movement of the oval regions within which the aurora occurs. Thus, on nights when active auroral storms are visible from mid-latitudes, the aurora may be in the equatorwards half of the sky, or even absent altogether from the sky at high latitudes.

The aurorae observed at mid-latitudes are, generally, less extensive than those of the high-latitude night. They do, however, share several common features. From locations such as Norway, at high latitudes in the auroral zone, displays may commence in early evening as quiet, often multiple, arcs which brighten after a time and become more active and extensive. Rayed forms exhibiting vigorous movement can fill the sky, and coronal aurora will often form. Later, weak arcs and patches of aurora are seen, as the observing site rotates with the Earth under the auroral oval from the evening to the morning sectors.

At high latitudes, such as those of north Norway, the observer is under the auroral oval for much of the time. Observers in mid-latitudes, however, will normally lie far from the oval, except under disturbed geomagnetic conditions. The geometry of the Earth's magnetic field and its interactions with the solar wind is such that the auroral ovals make their greatest equatorwards advance on the night-side, reflecting the distortion of the magnetosphere in the anti-solar direction. Consequently, the mid-latitude observer is brought closest to the auroral oval around *magnetic midnight*, about 22 hours local time for observers in the British Isles, for example. Long-term studies have, indeed, found magnetic midnight to be the time of peak activity in mid-latitude aurorae at a given location (Livesey, 1980).

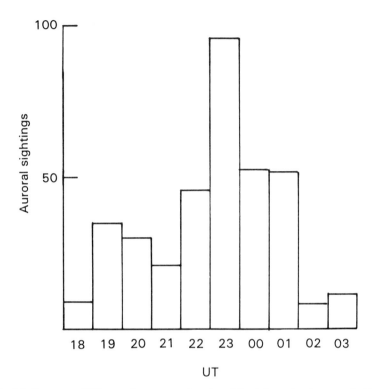

Fig. 7.8. Frequency of UK auroral reports as a function of local time, after data compiled by Ron Livesey, and used by kind permission. Aurorae are most frequently observed around the time of local magnetic midnight, when the observer is closest to the maximum equatorwards extent of the auroral oval.

REFERENCES

Dayton, L. (1989) Solar storms halt stock market as computers crash. *New Scientist* **123** (1681) 35.

Gavine, D. (1979) The Auroral Storm of 1978 May 1–2. *J. Brit. Astron. Assoc.* **89** (6) 607–610.

Gavine, D. (1995) Aurora and noctilucent clouds. In: Moore, P. (Ed.), *The observational amateur astronomer*. Springer.

Ham, R. (1991) Detecting solar radio waves. In: Moore, P. (Ed.), *1992 yearbook of astronomy*. Sidgwick & Jackson.

International Union of Geodesy and Geophysics (1963) *International Auroral Atlas* Edinburgh University Press.

Legrand, J. P., and Simon, P. A. (1988) Shock waves, solar streams and the spread of aurorae in latitude. *J. Brit. Astron. Assoc.* **98** (6) 311–312.

Livesey, R. J. (1980) The location of the polar aurora. *J. Brit. Astron. Assoc.* **90** (3) 253–262.

Livesey, R. J. (1985) The Aurora. *J. Brit. Astron. Assoc.* **95** (2) 67–69.

Livesey, R. J. (1988) *J. Brit. Astron. Assoc.* **98** (7) 335.

Livesey, R. J. (1989) The aurora 1987. *J. Brit. Astron. Assoc.* **99** (4) 172–178.

Livesey, R. J. (1990) The aurora 1988. *J. Brit. Astron. Assoc.* **100** (2) 73–78.

Livesey, R. J. (1994) The aurora 1993. *J. Brit. Astron. Assoc.* **104** (4) 183–188.

Livesey, R. J. (1995) The visibility of the aurora from the United Kingdom. *J. Brit. Astron. Assoc.* **105** (4) 179–181

Paton, J. (1973) The aurora. In: Moore, P. (Ed.), *Practical Amateur Astronomy.* Lutterworth.

Sampson, R. (1992) Fire in the sky. *Astronomy* **20** (3) 38–43.

Simmons, D. A. R. (1985) An introduction to the aurora. *Weather* **40** 147–155.

8

Other effects

8.1 SOLAR ACTIVITY AND UPPER ATMOSPHERIC EXTENT

As already discussed (section 1.2), the increased solar flux at X-ray and ultraviolet wavelengths around sunspot maximum leads to heating and expansion of the outer atmosphere. This can eventually cause artificial satellites to decay prematurely from orbit. The extent to which such effects are likely to operate is difficult to predict. Large flares can, however, produce a 10 000-fold increase in the solar X-ray flux, and may produce rapid changes in the extent of the upper atmosphere around the time of sunspot maximum.

The density of the atmosphere at 400–700 km altitude changes by a factor of 10 between sunspot minimum and sunspot maximum. The lower atmosphere shows a smaller degree of density change. Seasonal effects, with density maxima around the equinoxes (also seen to be favourable times for the occurrence of aurorae associated with high-latitude particle streams in the solar wind), are superimposed upon this gradual fluctuation. It is estimated that additional 'drag' resulting from expansion of the upper atmosphere associated with the Great Aurora of March 1989 reduced the orbital altitudes of some 6000 objects by about 1 km.

Observations of artificial satellites, particularly those in low orbits, can be of value in studies of upper atmospheric behaviour. Accurate timings of the passage of satellites across the celestial sphere can be used to assess how their orbits are evolving in response to changing atmospheric conditions. Such observations can profitably be obtained by amateur astronomers (Fea, 1973; Miles, 1985, 1995). Professionally operated Hewitt cameras allow extremely precise time-position data for satellites to be obtained.

8.2 GEOMAGNETIC EFFECTS

In addition to producing the visual spectacle of the aurora, the arrival in near-Earth space of energetic electrons in the solar wind disturbs the geomagnetic field. This disturbance can be measured using appropriate *magnetometer* equipment. Disturbances of the terrestrial magnetic field in association with auroral effects were noted as early as the eighteenth century, using simple equipment (section 3.3).

The magnetic field at ground level may be described by three principal characteristics. These are D, the *angular deviation* (corresponding to the east–west swing of the compass needle relative to the magnetic pole which it seeks under undisturbed conditions), *H*, the *horizontal field strength* and *Z*, the *vertical field strength*. These characteristics all vary under disturbed geomagnetic conditions which may be accompanied by auroral activity. Field strengths are expressed in nanoteslas (nT), equivalent to the geophysicists' unit, the gamma: 1 nT = 0.000 01 gauss.

8.2.1 Magnetometers

The variations of *H*, *Z* and *D* at a given location can be measured using a magnetometer. Much of the environment in built-up areas is magnetically 'noisy' as a result of human activities. Professional observatories in quiet locations, however, can very sensitively measure the fine-scale fluctuations of the local magnetic field. Results are usually displayed on a *magnetogram*, a plot of field strength or direction versus time. The magnetogram trace is indicative of the nature of any disturbance.

In its simplest form, the magnetometer need only be a suspended, free-swinging magnet which will respond to the changing local magnetic field by re-orienting itself. If these swings can be measured and recorded, the observer has a means of detecting fluctuations

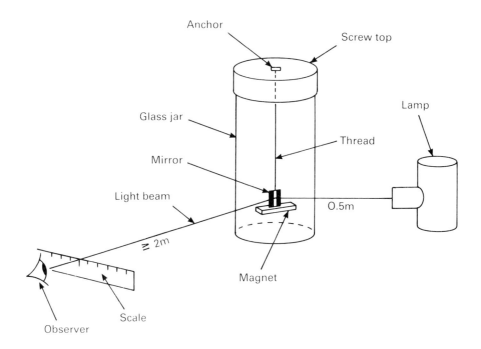

Fig. 8.1. The simple, but effective, jam-jar magnetometer allows fluctuations in the horizontal component of the Earth's magnetic field to be measured, and affords the possibility of obtaining early warning of auroral activity. The free-swinging bar magnet, protected from draughts by being enclosed in a clear container, aligns itself to the magnetic field. Fluctuations can be sensitively detected by regularly measuring the position of the reflected light beam from a mirror mounted on the magnet. Several of these devices, constructed to this design, are operated by amateur astronomers and others across the world. Drawing courtesy of the jam-jar magnetometer's designer, Ron Livesey.

in the value of *D*, the angular deviation of the field. Such a device is the *'jam-jar' magne-tometer* devised by Ron Livesey of the BAA Aurora Section in the early 1980s (Livesey, 1982, 1989).

The jam-jar magnetometer comprises a reasonably powerful suspended bar magnet, protected from draughts by being hung on its thread inside a clear jar: despite the device's popular name, most users have found screw-top instant coffee containers best. The thread from which the magnet is suspended passes through a small hole in the centre of the con-tainer's lid. A small strip of mirror is attached to the magnet, and onto this is shone a reasonably narrow (crudely 'collimated') beam of light from a torch or other convenient source from a distance of, preferably, a few metres. The position of the reflected spot of light from the mirror on the opposite side of the room, garage, or other convenient location for the magnetometer, is measured against a suitable scale such as a 30-cm ruler. Use of a container of reasonable optical quality helps in sharpening the light beam and its subse-quent reflection.

Readings taken at intervals will reveal the diurnal variation (section 8.2.2) and the oc-currence of any—possibly aurora-related—magnetic activity. Operators of the jam-jar magnetometer system have picked up, clearly, crotchets, Sudden Storm Commencements and storm/substorm activity. Early warnings of possible visible aurora have been issued to amateur observers in Scotland on the basis of magnetic field activity detected by jam-jar magnetometer operators.

It is, sometimes, the case that the disturbances detected using these simple but effective devices do not coincide with aurora visible at the latitude of the magnetometer operator. Apparently, magnetic disturbances at ground level can propagate to lower latitudes than the visibility of the atmospheric displays which cause them. In some respects, magnetome-try of this kind offers those at lower latitudes (as in southern England, or the mid-United States) their most reliable and frequent means of following the effects of solar–terrestrial interactions.

An important factor in successful operation of the simple jam-jar magnetometer is that it be located in a 'quiet' environment. Spare rooms or garages are favoured locations, but it seems that movement of metallic objects (garden implements in a garage, for instance) can alter the device's local magnetic field by a measurable extent! Likewise, the sensitivity of a well-made jam-jar magnetometer is sufficiently high that it can detect magnetic fluc-tuations due to passing motor cars in the vicinity.

Once a good location has been found, the set-up should not be moved, and results there-after will be consistent from day to day. While many operators record the deviations from jam-jar magnetometers by eye, it is possible, with a little ingenuity, to construct electroni-cally recording versions. It is worth noting that the best jam-jar magnetometer records may be quite readily aligned with magnetograms recorded at professional magnetic observato-ries using much more sophisticated equipment.

Fluctuations in the orientation of the magnet in jam-jar systems can frequently be corre-lated with interludes of increased activity during auroral events, as has been demonstrated on several occasions when visual and magnetometer observations have been obtained si-multaneously.

Slightly more complicated than the jam-jar system is the *magneto-resistive magne-tometer* designed and used with great success by the late Doug Smillie, who was an active

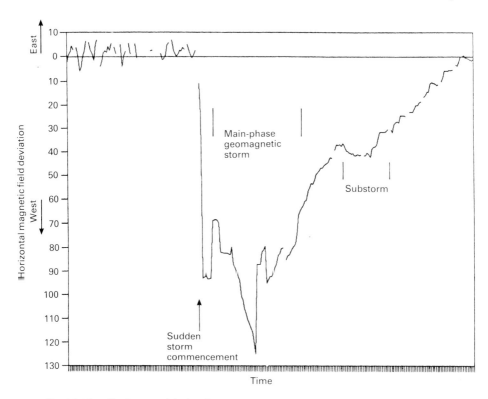

Fig. 8.2. The effectiveness of the jam-jar magnetometer in recording magnetic field fluctuations is clearly shown in this magnetogram, reproduced by kind permission of the observer, Tony Michell of Beckenham, Kent, England. Mr Michell uses a personal computer to record the motions of the free-swinging magnet. This magnetogram shows the progress of a geomagnetic storm around December 24–25 1990, resulting from flare activity associated with a major sunspot group: poor weather prevented the visible auroral display from being very widely seen in the British Isles. At left, minor fluctuations of the undisturbed field may be seen. Arrival of the shock wave from the flare in the solar wind produces the sudden storm commencement (SSC), indicated by the arrow. The main-phase storm then set in, followed by a gradual recovery phase, on which a substorm can be seen superimposed towards the right of the magnetogram. Recovery of the magnetic field to its quiet condition took about three days from the SSC.

contributor to the BAA Aurora Section based in Wishaw, Scotland (Smillie, 1992). This, again, uses a free-swinging bar magnet, suspended in a partially oil-filled container. Magneto-resistive (Hall-effect) diode sensors located close to the poles of the magnet on the outside of the container allow both the horizontal and vertical motions of the magnet in response in response to changing magnetic field conditions to be measured. The signal output is usually to a chart recorder, and the resulting magnetograms testify to the excellent sensitivity of the instrument.

Another alternative, and very sensitive, system has been designed and constructed by the Swedish amateur astronomer Gote Flodqvist (Flodqvist, 1993), using a phototransistor to detect the movement of a compass needle across the beam from a light-emitting diode.

More complex in its working principle and construction, but still accessible to the amateur worker in possession of electronics skills, is the *fluxgate magnetometer* (Pettitt,

1984), which can very sensitively detect fluctuations in the local magnetic field *strength*. Since they measure field strength rather than deviation, fluxgate systems are, apparently, better able to detect Sudden Storm Commencements than the swinging-magnet types.

8.2.2 Behaviour of the magnetic field

Even under quiet solar-geomagnetic conditions, when auroral activity is confined to high latitudes, fluctuations in the magnetic field at mid-latitudes can still be detected. The Sun and Moon raise tides in the ionosphere, causing it to rise and fall slightly. Consequently, the strength of the ground-level magnetic field induced by electric currents in the ionosphere is subject to a gradual *diurnal variation*.

Increased solar activity produces a number of detectable effects. The transient increases in D-region ionization produced by solar flare-emitted X-rays which cause Sudden Ionospheric Disturbance events (section 6.1.2) are responsible for magnetic *crotchets*. During SIDs, the ionospheric conductivity rises, with consequent increases in the induced ground-level magnetic field. In common with radio SIDs, crotchets are detected only on the dayside of the Earth.

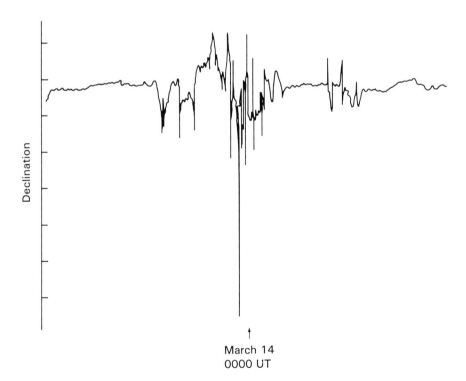

March 14
0000 UT

Fig. 8.3. Magnetogram showing the very violent fluctuations in magnetic field direction and strength during the Great Auroral Storm of March 1989. The most extreme deviation of the magnetic field in the British Isles was recorded around 22 hours UT, at the peak of the visible event.

The fluctuations produced in crotchet events give a fairly regular 'saw-tooth' trace on a magnetogram. Less regular are the traces produced when the main concentration of particles ejected by a solar flare arrives in near-Earth space via the solar wind. Compression of the magnetosphere by the supersonic shock-wave at the leading edge of a flare ejection on the solar wind briefly intensifies the Earth's magnetic field, and is recorded as *Sudden Storm Commencement* (SSC; section 5.4.2.2). SSC events are often, but not always, followed by the development of auroral activity extending to lower latitudes.

Where SSC is followed by extensive auroral activity, the geomagnetic field is also found to be active, during the *main-phase storm*, which can last for many hours. Wild fluctuations may occur during the main-phase storm, whose effects can be particularly marked locally in the period during which the magnetic observatory is closest to the disturbed auroral oval, on the Earth's night-side. Peaks of activity during which the aurora becomes coronal give rise to very sharp fluctuations in local magnetic field intensity (bringing a virtual 'collapse' in severe geomagnetic storms). The main-phase storm is brought on by intense currents flowing in the highly ionized upper atmosphere, as electrons rain in from the magnetotail plasma sheet.

The *recovery phase* following the onset of a geomagnetic storm can take several days, particularly after a major solar flare event. Superimposed on the gradual recovery are *bays*, corresponding to substorm activity: under normal conditions, only those observatories at higher latitudes will detect the magnetic field variations due to substorms.

8.2.3 Indices of geomagnetic activity

Magnetometers are operated by numerous professional bodies for the purposes of monitoring the temporal fluctuations of the global magnetic field. Indices of daily activity are published by a number of these institutions, and can subsequently be correlated with auroral activity. Several different indices are published (Simon, 1978); a few of the more commonly used are listed below.

The aa index
One useful method of measuring the degree of disturbance of the geomagnetic field is to compare activity at two stations at equivalent—antipodal—locations in either hemisphere. Such a method has been employed since 1868 by the Institut de Physique du Globe de Paris, to produce the aa index. The aa index is based on 12-hour averages of activity from antipodal stations. As estimated on this scale, the geomagnetic storm associated with the Great Aurora of March 1989 (section 2.7) had an aa index of 462: for comparison, the aa index reached 546 during the September 1909 event.

The Kp index
The University of Gottingen is the location of one of the world's earliest-established magnetic observatories. Since 1932, the Institut für Geophysik at Gottingen has originated what is probably the the most widely used index of geomagnetic activity, the Kp index. This index is based on magnetometry by 12 observatories worldwide, mostly at sub-auroral locations in the northern hemisphere. These record variations in field intensity, which are subsequently averaged for all stations over intervals of 3 hours. Kp is determined on a

semi-logarithmic scale, based on the average variation in magnetic field intensity as indicated in Table 8.1.

Table 8.1. The Kp Index

Variation (nT = gammas)	0	5	10	20	40	70	120	200	330	>500
Kp index		0	1	2	3	4	5	6	7	8 9
3-hour *ap* equivalent		0	3	7	15	27	48	80	140	240 400

A planetary index, Kp, of greater than 5 implies storm conditions, and can often be correlated with visible displays at lower latitudes. Summaries of the daily Kp indices for the previous month are issued monthly. Each month's summary is accompanied by a longer—four solar rotations—'music-note' diagram of the Kp in graphic form, aligned as a Bartels-type 27-day summary chart, allowing recurrent events to be seen and, possibly, forecast.

The Ap index
An alternative index of geomagnetic activity, again originated from Gottingen, is the Ap index. This is similar to the aa index, in taking account of readings taken at antipodal stations, but averaged over 24-hour intervals. Unlike the roughly logarithmic Kp index, Ap is a *linear* measure of magnetometer deviation. The two indices can, however, be broadly related by the 3-hourly equivalent *ap index* (Table 8.1).

8.3 FORBUSH DECREASES AND GROUND LEVEL ENHANCEMENTS

The Galaxy, in one of whose spiral arms our solar system is located, is permeated by cosmic rays—energetic particles, whose movements are to some extent governed by the galactic magnetic field. These particles, chiefly protons, are accelerated to high energies during extremely violent events such as supernova explosions. Similar particles in the lower solar atmosphere are accelerated during solar flare events, but these do not generally reach the extreme energies attained by galactic cosmic ray particles.

Galactic cosmic rays entering the heliosphere (section 4.3.9), are deflected by the heliospheric magnetic field, limiting their access to the inner solar system.

Cosmic rays arriving at Earth may be detected in a number of ways, including the use of Geiger counters, scintillation counters, cloud chambers, or photographic emulsions (Friedlander, 1989). Detectors have been flown in balloons at high altitudes, and carried aboard satellites. Detection at ground level is dependent on the arrival of secondary products resulting from collisions between the primary cosmic ray and constituents of the atmosphere. The atmosphere therefore plays an important role in protecting the Earth's surface, and living organisms thereupon, from continual bombardment by these extremely energetic particles.

Geiger counters at the Earth's surface detect the arrival of electron air showers, resulting from the secondary 'cascade' produced on the arrival of a highly energetic primary cosmic ray. Detectors at high altitude observatories (for example, Calgary, Canada, at an altitude of 1100 metres) can be used to measure the flux of neutrons.

At times of high sunspot activity, the magnetic field of the heliosphere is more turbulent, and becomes more effective in retarding the penetration of galactic cosmic rays to the inner solar system. The galactic cosmic ray flux is therefore seen to be the inverse of the sunspot cycle: the average galactic cosmic ray frequency measured by detectors on Earth reaches its maximum in the years around sunspot minimum.

Around sunspot maximum there are interludes during which the galactic cosmic ray flux may diminish still further, as a consequence of solar flare activity. At these times, high-speed shock-waves ploughing outwards through the solar wind act as a further barrier to incoming galactic cosmic rays. These shock-waves have been visualized as coronal transients using spacecraft-borne coronagraphs (section 4.3.4).

These interludes of decreased galactic cosmic ray flux were first noted in detail by Scott E. Forbush and his colleagues in 1942, and are now known as *Forbush decreases*. Instruments aboard the Pioneers and Voyagers have shown that flare-associated shock-waves can give rise to Forbush decreases at least as far as 20 AU from the Sun (Sakurai, 1987). A major Forbush decrease in 1991 was detected at 53 AU from the Sun by the Pioneer spacecraft.

The onset of a Forbush decrease at Earth is coincident with the arrival of the shock front in the solar wind, marked in geomagnetic activity by Sudden Storm Commencement (section 8.2.3).

While the flux of galactic cosmic rays is reduced during Forbush decreases, these events may also be accompanied by an increased flux of cosmic rays of solar origin, effectively masking the Forbush decrease (Schatten, 1990). Solar particles can become accelerated to very high energies (GeV) in the kernel of the parent solar flare. Arrival of large numbers of such solar cosmic rays produces an increase in the neutron flux, detected as a *Ground Level Enhancement* (GLE). Several GLEs were recorded around the vigorous sunspot maximum of 1989–90 (Mathews and Venkatesan, 1990).

Relatively unconstrained by magnetic fields, particles accelerated to high energies can arrive in the near-Earth neighbourhood within a few hours of solar flares, ahead of those which give rise to visible auroral effects. For example, Mathews and Venkatesan report a significant GLE on October 19, 1989, which was not followed by auroral activity until October 21–22 (when coronal forms were visible from higher mid-latitudes, and aurora was seen to southern England).

The particles which give rise to GLEs are, potentially, hazardous for astronauts or cosmonauts aboard spacecraft, above the shielding atmosphere. Energetic cosmic ray particles passing through tissue can cause several forms of damage to cells, including the creation of free radicals, and damage to DNA. In healthy individuals, the latter damage is normally dealt with quite rapidly. Accumulation of DNA damage, resulting from (genetically inherited) defects in metabolic repair pathways, has been implicated as a cause of several forms of cancer. The effects can also be dose-dependent: major solar flare particle releases will have more serious effects.

8.4 RADIO AURORAE; IONOSPHERIC EFFECTS

As viewed from large distances above the Earth, the aurora is a strong emitter of radio waves in the frequency range from below 100 kHz to nearly 1 MHz, peaking around 300 kHz. This naturally occurring *Auroral Kilometric Radiation* (AKR) is stronger than any man-made transmissions at such frequencies, averaging 10^7 watts, and sometimes reaching peaks of 10^9 watts. The Earth's AKR is not detectable by receivers on the ground, however, being reflected upwards by the ionospheric E-layer at 110 km altitude. Nonetheless, auroral activity does have important consequences for radio communication, and can be both problematical and beneficial for those attempting long-distance communications.

Sudden Ionospheric Disturbances (section 6.1.2) resulting from solar flares are detrimental to HF radio communication. These events lead to fadeouts as a result of increased D-region ionization and absorption which occurs over the whole day-side of the Earth, irrespective of geomagnetic latitude. At high latitudes, the arrival of accelerated protons from a solar flare event can give rise to Polar Cap Absorption events (section 6.4.2), during which HF radio communication is disrupted. Enhanced ionization at the altitudes of the higher ionospheric E- and F-layers can also be disruptive, again as a result of increased absorption.

Enhanced ionization under auroral conditions can also be used for the *reflection* of VHF radio waves. Reflection perpendicular to auroral arc structures comes directly back to the transmitter as *backscatter*. Professional ionospheric research often involves the use of powerful radar signals (section 3.9), which may be detected as backscatter from auroral structures using amateur radio equipment (for example, a 2-metre, 5-element horizontally polarized Yagi aerial). Backscatter from transmitters not normally receivable at a given location can give forewarning of auroral conditions.

Anomalous reception of signals from radio *beacons* can also indicate the onset of auroral conditions. Typically, such beacons have a range of about 300 km in all directions under normal conditions. Reflection of beacon signals from an auroral arc can alter the range in certain directions. For example, the Black Isle Beacon 'went auroral' (that is, became receivable!) for radio operators in central Scotland, for whom it is normally too distant for reception, during the March 1989 storm.

Rather than aiming to detect backscatter from research transmitters or beacons, amateur radio operators using the universal 144.3 and 433 MHz calling channels usually prefer to 'work' the aurora in order to obtain longer-than-normal distance (DX) contacts with other operators. Basically, the aurora can be used as a radio-reflective surface from which VHF signals can be bounced, to permit 'bi-static' backscatter contacts. The electron densities required to permit these contacts are very much higher than those needed to give rise to visual aurorae. Radio auroral contact depends on *free* electrons, stripped from their parent atomic nuclei. Visual and radio events need not overlap; for instance, radio amateurs enjoyed excellent auroral contact conditions in February 1984, at a time when low-latitude visual events were scarce.

From a given location, there is a unique, theoretical limit—the 'boundary fence'—within which bi-static auroral contacts are possible. This typically defines an oval region extending some 2000 km east–west, and 1000 km in the north–south direction (Newton, 1991). The best reflection is from discrete auroral arcs aligned to lines of geomagnetic latitude. Operators in the UK, for instance, use discrete arcs lying north of Shetland for the reflection of signals eastwards into Scandinavia, or to eastern European countries such as Hungary. The unusual situation of auroral activity *south* of the UK allowed British radio amateurs to make contacts with operators in Italy during the Great Aurora of March 1989.

Movement of, and turbulence within, the reflective auroral surface leads to interference and signal distortion. Voice and Morse transmissions become 'raspy' (and, often, difficult to decipher) as a result of Doppler motions, and interference as successive radio waves overlap. Similar interference phenomena experienced by 420-line television receivers allowed a degree of early warning for aurora observers in the International Geophysical Year period (section 3.8); more modern systems are less subject to such interference.

Fig. 8.4. A typical amateur radio operator's aerial equipment. Signals scattered from auroral ionization may be detected using such aerials. The operators at this particular station—Chris and Ken Sheldon in Worcestershire, England—can sometimes obtain early warning of auroral conditions as a result of enhanced radio signal propagation.

Enhanced signal reflection during auroral conditions is primarily from the clouds of ionization in the aurora itself, rather than from enhancements of the ionosphere.

Both time of day and geomagnetic activity influence the efficiency of auroral radio contacts. Under quiet geomagnetic conditions (Kp <5), the peak period for establishing long-distance contacts from a mid-latitude station is between 17 and 19 hours local time, with a secondary peak around 21 hours (Lange-Hesse, 1967). Around local magnetic midnight, when the station lies closest to the Harang discontinuity (section 5.3), auroral radio communication conditions are very poor.

Seasonal effects also operate: the secondary peak is more pronounced around the equinoxes. Summer is generally a poor time for radio aurora DX contacts. Geomagnetic storm conditions (Kp>5) also enhance the secondary peak in auroral backscatter bi-static communications.

The efficiency of auroral radio contacts also varies as a function of time in the sunspot cycle. In contrast with visual events, radio aurorae show a peak after sunspot maximum. In sunspot cycle 21, for example, the peak time for bi-static contacts was in 1982, some years after the most vigorous low-latitude visual events occurred. This peak corresponds more closely to the coronal hole 'season' in the declining phase of the solar cycle than to the flare-associated events close to sunspot maximum, though the latter do give rise to enhanced communication conditions on occasion.

Communications can occur only if the auroral oval lies within a certain range of the transmitting and receiving stations. Such conditions are met around the times of the observed peaks. Onset of increased geomagnetic activity and expansion of the oval (section 6.3.1) will alter the peak times for bi-static contacts, which are then made earlier and later: typically, under conditions of Kp >5, peak times for long-distance auroral contacts are 14–19 hours and 21–00 hours local time.

Contact during intense storms can be very erratic due to rapid movements within the aurora, and the development of denser clouds of ionization which may absorb even VHF signals.

Evidence for the possible 'flash' aurorae (section 7.2.3) has been found by amateur radio operators, who are occasionally able to establish contacts which have auroral characteristics (interference and Doppler-motion distortions) lasting for only very brief periods.

Amateur radio operators are primarily interested in auroral phenomena as a means of attaining long-distance contacts. It is a tradition to exchange and collect call-signs, and address cards, after a contact has been made. Some radio 'hams' aim, particularly, to beat personal distance records, which depend heavily on the prevailing auroral conditions. While the 2-metre band is most widely used, some amateur radio operators have also made use of the 6-metre band (which requires lower electron densities) to obtain trans-Atlantic auroral contacts.

Radio auroral phenomena can occur in the absence of visual events, and vice versa. Radio events seem more frequently to be recurrent, and are thus somewhat easier to forecast. Summaries of recent activity and contacts are published, for example, by the Radio Society of Great Britain (RSGB) in their monthly journal *Radio Communication* (RADCOM to the aficionados), along with forecasts for possible activity in the immediate future. Similar reports and forecasts appear in the monthly magazine *Practical Wireless*.

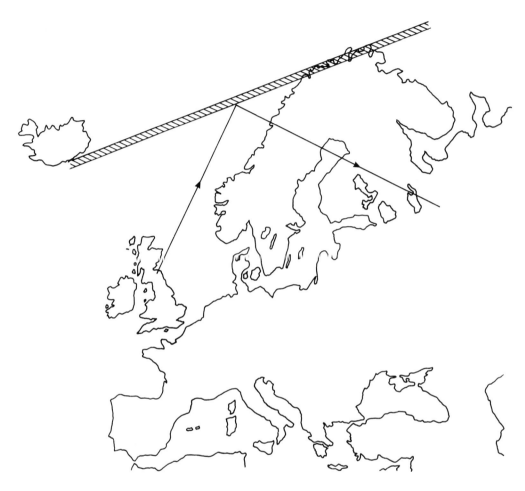

Fig. 8.5. Use of an auroral arc to scatter radio signals over greater than normal distances. In this instance, an arc—aligned to the geomagnetic field lines, and lying some way north of the transmitter—is used to reflect radio signals into eastern Europe from Britain

American radio amateurs have access to similar facilities via the American Radio Relay League. Forecasts of likely activity are also broadcast on amateur channels.

Less regular or predictable is another form of ionization which occurs at auroral altitudes, but is not apparently connected to auroral phenomena. This *Sporadic E* is found most frequently at mid-day during the summer, at mid-latitudes. Sporadic E ionization appears in thin sheets at about 100 km altitude, and affects areas of 1000–2000 km. It manifests by producing interference in radio and other transmissions: for example, Sporadic E is the cause of interference between television signals from transmitters in the UK, and in continental Europe, during the summer.

The mechanism by which Sporadic E arises remains a subject for investigation. It has been proposed that wind-shear and magnetic field effects can combine to locally intensify existing ionization in the high atmosphere (Hargreaves, 1979). It has also been pointed out, by some observers, that Sporadic E occurs most frequently at those times when noctilucent clouds (section 9.1) are common at high temperate latitudes. While the broader latitudinal distribution of Sporadic E makes it unlikely that a causative relationship exists between the two phenomena, it is perhaps of interest to note that both have shown a similar apparent secular increase in frequency of occurrence into the latter parts of the twentieth century (section 9.1.2).

8.5 THE AURORAL SOUND CONTROVERSY

As a phenomenon of the high atmosphere, the aurora should, sensibly, be so far distant from the observer that no sound can be heard. Several witnesses, however, are adamant that they have heard sound during times of intense auroral activity, occurring simultaneously with visual outbursts—often as arcs or bands pass overhead during the breakup phase of substorm displays, for example (section 6.4.3). The interval during which sounds are heard by such witnesses, who have included professional scientists engaged in auroral research at high-latitude sites, is typically short, perhaps of the order of 10 minutes.

Descriptions of the sounds heard include faint whistling, rustling, swishing and crackling or a soft hissing. Similar sounds are sometimes reported in association with meteors. As with auroral sounds, reports of simultaneous sounds from meteors are, in general, greeted with considerable scepticism by the scientific community.

There are two obvious problems with simultaneous auroral (or meteor) sound. Firstly, the aurora occurs in near-vacuum, mostly in the tenuous upper atmosphere above 100 km altitude, such that there is very little medium through which sound waves can be propagated downwards. Perhaps more critically, the speed of sound in the atmosphere is such that any noises produced at auroral altitudes should take a matter of minutes to reach ground level, making it hard to reconcile the occurrence of auroral sounds simultaneous with visual activity.

A number of explanations have been offered to account for the possibility of auroral sound. One, physiologically based, suggestion is that the sounds are simply a consequence of the ears' adjustment to cold outdoor conditions. This does not, however, account for the occasions on which witnesses report *no* sounds under conditions of high activity. Alternatively, the effect may merely be psychological, with observers subconsciously *expecting* to hear sounds associated with vigorous activity.

Among the physical mechanisms proposed, propagation of VLF radio waves, and their more-or-less instantaneous re-transmission from suitable nearby 'receivers' has enjoyed some support, both for auroral and meteor sounds. A more detailed investigation of VLF propagation from very bright meteors (fireballs) by S. M. Silverman and T. F. Yuan and colleagues, however, has cast doubt on this electrophonic mechanism (Wang *et al.* 1984). An alternative is *coronal discharge* from pointed objects (such as the tips of coniferous trees) in the vicinity of the observer. Coronal discharge requires the generation of strong electrical fields, which may be correlated with auroral activity. This mechanism is dependent on the dryness of the air, perhaps offering an explanation for the reporting of simultaneous sound associated with some auroral displays, but not others.

While hundreds of reports of anomalous auroral sound have been made since the eighteenth century, it is perhaps significant that attempts to record these using modern equipment have produced only ambiguous results. The occurrence or otherwise of auroral sound is likely to remain contentious for some time to come.

8.6 LUNAR ECLIPSE BRIGHTNESS AND THE SOLAR CYCLE

On occasions when the alignment of the nodes of its orbit are appropriately positioned, the Moon may undergo eclipse in the Earth's shadow at opposition to the Sun in the sky. Eclipses may be total or partial. Studies on the brightness of total lunar eclipses suggest that those which occur at the beginning of a sunspot cycle are darker than those at a cycle's end (Shepherd, 1982). In particular, an eclipse at sunspot minimum on 1964 June 25 appears to have been particularly dark. The correlation of lunar eclipse brightness appears to be with mean heliocentric latitude of sunspot groups, rather than with overall levels of solar activity.

The influence of the dust-load in Earth's atmosphere may, however, be more significant. A well-observed total lunar eclipse in December 1992 (two years after sunspot maximum) was notably dark, presumably as a result of volcanic dust from the previous year's Mount Pinatubo eruption suspended in the high atmosphere. Such dust limits the amount of sunlight refracted around the dark body of the Earth into the shadow cone.

It has been suggested that high-energy protons and electrons from the solar wind may be deflected by Earth's gravitational and magnetic fields to impinge on the Moon's surface at the times of total lunar eclipses. Meteoritic material, which constitutes an important part of the lunar surface, has been found to emit light at 670 nm wavelength (in the red part of the spectrum) under bombardment with protons in laboratory conditions. A similar effect has been invoked to account for the unusual brightness of the Moon during some eclipses. Localized effects of this nature may also be a partial explanation of the enigmatic Transient Lunar Phenomena occasionally reported by observers of the Moon.

Bombardment of the lunar surface by solar wind particles is responsible, in part, for creating the Moon's extremely tenuous atmosphere. A neutral sodium cloud raised by these effects streams away down the solar wind from the Moon (Stern, 1993).

REFERENCES

Fea, K. (1973) Earth satellite observation by amateurs. In: Moore, P. (Ed.), *Practical amateur astronomy*. Lutterworth.

Flodqvist, G. (1993) Detecting the polar lights. *Sky and Telescope* **86** (4) 85–87.

Friedlander, M. W. (1989) *Cosmic rays*. Harvard University Press.

Hargreaves, J. K. (1979) *The upper atmosphere and solar-terrestrial relations*. Van Nostrand Reinhold.

Lange-Hesse, G. (1967) Radio aurora. In: McCormac, B. M. (Ed.), *Aurora and airglow*. Reinhold.

Livesey, R. J. (1982) A 'jamjar' magnetometer. *J. Brit. Astron. Assoc.* **93** 17–19.

Livesey, R. J. (1989) A jam-jar magnetometer as 'aurora detector'. *Sky and Telescope* **78** 426–432.

Mathews, T., and Venkatesan, D. (1990) Unique series of increases in cosmic-ray intensity due to solar flares. *Nature* **345** 600–602.

Miles, H. (1985) Artificial satellites. *J. Brit. Astron. Assoc.* **95** 164–166.

Miles, H. (1995) Artificial satellites. In: Moore, P. (Ed.) *The observational amateur astronomer.* Springer.

Newton, C. (1991) *Radio auroras.* Radio Society of Great Britain.

Pettitt, D. O. (1984) A fluxgate magnetometer. *J. Brit. Astron. Assoc.* **94** 55–61.

Sakurai, K. (1987) Cosmic rays and energetic particles in the heliosphere. In: Akasofu, S.-I., and Kamide, Y. (Eds), *The solar wind and the Earth.* D. Reidel.

Schatten, K. H. (1990) The Sun's disturbing behaviour. *Nature* **345** 578–579.

Shepherd, J. (1982) Lunar eclipses, lunar luminescence and transient lunar phenomena. *J. Brit. Astron. Assoc.* **92** 66–68.

Simon, P. (1978) Activite solaire, indices et catalogues. *L'astronomie* **62** (2) 61.

Smillie, D. J. (1992) Magnetic and radio detection of aurorae. *J. Brit. Astron. Assoc.* **102** (1) 16–20.

Stern, A. (1993) Where the lunar winds blow free. *Astronomy* **21** (11) 36–41.

Wang, D. Y., Tuan, T. F., and Silverman, S. M. (1984) A note on anomalous sounds from meteor fireballs and aurorae. *J. Royal. Astron. Soc. Canada* **78** (4) 145–149.

9

Related high-atmosphere phenomena

9.1 NOCTILUCENT CLOUDS

9.1.1 The origin and nature of noctilucent clouds

At those latitudes most favoured for observations of aurorae, the summer night-time sky never becomes completely dark. In northern Scotland at midsummer (June 21st), for example, the Sun never sinks more than about nine degrees below the northern horizon. Under such circumstances, twilight persists all night, and only the brighter naked-eye stars are readily visible: astronomical observations are difficult, and this might well be regarded as 'off-season' for aurorae, which will normally be swamped by the bright sky background. There is, however, one phenomenon of the high atmosphere, whose occurrence is thought to be intimately influenced by auroral conditions, and of which observations can usefully be made during the twilit summer nights at higher latitudes: *noctilucent clouds*.

It may seem unusual for those normally engaged in astronomical work to actively look for cloud phenomena, but noctilucent clouds are quite different from the lower, tropospheric, clouds which cause so much frustration to observers on the ground. While the highest tropospheric clouds (the 'mares' tails of cirrus) reach a maximum altitude of about 15 km in the atmosphere, noctilucent clouds are formed in thin sheets at altitudes in excess of 80 km. Since there is very little water vapour present in the atmosphere at such great heights, noctilucent clouds are extremely tenuous, and may be seen only under certain illumination conditions; they are lost against the bright daytime sky, only becoming visible as twilight deepens after sunset. In order for noctilucent clouds to become visible, the Sun must lie between 6 and 16 degrees below the observer's horizon: too high, and the sky will be over-bright, too low and the noctilucent clouds themselves will be in the Earth's shadow. Fig. 9.1 illustrates the illumination conditions necessary for noctilucent clouds to be observable. The sky must be sufficiently dark for the noctilucent clouds to appear bright by contrast, while the clouds themselves remain in sunlight above the Earth's shadow. Looking up through the Earth's shadow, the observer should see any foreground tropospheric clouds in darkness against the noctilucent cloud field over the poleward horizon.

Noctilucent clouds are believed to form only during the summer months in either hemisphere. The noctilucent cloud 'observing season' in northwest Europe and Canada extends through June and July (perhaps peaking in late June and the first week of July), while

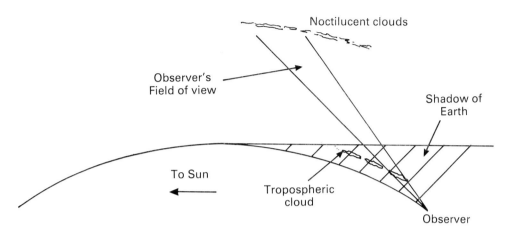

Fig. 9.1. Noctilucent clouds become visible at high temperate latitudes during the summer months by virtue of their great height (in excess of 80 km). Noctilucent clouds are sufficiently high to remain in sunlight, after tropospheric clouds are in darkness in the Earth's shadow. In this figure, the curvature of the Earth is exaggerated.

southern hemisphere observers are favoured in December and January. Records of noctilucent cloud sightings in the southern hemisphere are rather sparse by comparison with those from the northern hemisphere, thanks mainly to a less favourable disposition of the main land-masses there relative to the main zones of occurrence. Noctilucent clouds usually form between latitudes of about 60 and 80 degrees, and may be observed to as low as about 50 degrees latitude. In contrast with the aurora, the regions of most frequent noctilucent cloud occurrence are determined principally by geographical latitude. Observations are most frequently made from those latitudes where the summer night sky is reasonably dark, and the cloud-sheet may be observed at an acute angle; Aberdeen (57°N) and Helsinki (61°N) are favoured in this respect, for example.

During the spring and summer months at high latitudes, upwelling of cold, moist polar air is thought to carry water vapour to the extremely high altitude of the mesopause. Average temperatures at the mesopause are low, of the order of 165 K (−108°C). Temperatures as low as 111 K (−162°C) have been measured during rocket flights into noctilucent cloud fields (Gadsden, 1989).

At the mesopause, water vapour may condense around small nuclei to form noctilucent clouds. The precise nature of the condensation nuclei remains subject to debate, despite attempts to recover noctilucent cloud particles using 'Venus Flytrap' sounding rockets over Sweden in the 1960s (Soberman, 1963), and further rocket flights into noctilucent clouds over Sweden and Canada up to the early 1970s. Direct examination of noctilucent cloud material by spectroscopic techniques from ground level is rendered impossible by the bulk of atmosphere between observer and target.

A logical source of condensation nuclei might be meteoric debris: meteors become luminous at heights of about 100 km above the Earth's surface, and residual particles from each meteor's ablation should remain suspended in the high atmosphere for periods estimated at about three years. Volcanic material injected into the stratosphere during violent events such as the Krakatoa eruption in 1883 or Mount Pinatubo in 1991 may also be carried aloft to provide a source of condensation nuclei (and, possibly, water vapour) for noctilucent clouds.

Historically, noctilucent clouds were first observed in the years immediately following the Krakatoa explosion, perhaps lending credibility to the possibility of volcanic sources of nuclei. It has, however, also been suggested that the increased prevalence of spectacular sunset and twilight optical phenomena following the global dispersion of the dust ejected by the Krakatoa explosion may have led to improved observational vigilance at those times of night when noctilucent clouds might be expected to become visible, thereby accounting for their initial detection in the mid-1880s (Gadsden and Schroder, 1989).

Photoionization of atmospheric particles by solar ultraviolet radiation may provide another possible source of condensation nuclei for noctilucent cloud particles.

The source of the water vapour widely believed to comprise the bulk of noctilucent clouds is also subject to debate. One theory suggests that methane released into the atmosphere by industrial activities such as oil exploration and coal mining may rise into the stratosphere, there to be dissociated by solar ultraviolet radiation above the ozone layer with the resulting formation of two molecules of water per methane molecule as a by-product (Thomas *et al*. 1989). Carried higher into the atmosphere, the water could condense to form noctilucent clouds. This theory may also account for an apparent dearth of noctilucent cloud sightings in records prior to the late nineteenth century there previously having been insufficient quantities of water vapour present in the high atmosphere to allow cloud formation. Increasing liberation of methane with time has been postulated as leading to an increased frequency of noctilucent cloud displays, and a parallel increase in the average brightness of those displays. Some observations from northwest Europe suggest that the northern hemisphere noctilucent cloud observing season may have lengthened during the late 1980s, with sightings being made earlier into May and later into August than previously. Further observations are required in order to assess the reality of these suspected changes.

Fig. 9.2 summarizes total numbers of noctilucent cloud sightings from northwest Europe over a 24-year period. The gradual long-term increase in noctilucent cloud frequency is apparent, as are superimposed decreases in frequency which appear to be correlated with periods of high sunspot activity in 1970 and 1980.

Condensation of water vapour around the nuclei, whatever their nature, appears to begin at an altitude of about 85 km. As the particles grow in size and mass, they begin to fall under gravity, reaching a maximum density about 82 km altitude. Below this level, the atmospheric temperature begins to rise, and the water evaporates. Consequently, noctilucent clouds appear in thin sheets.

While noctilucent cloud sheets are, by the standards of tropospheric clouds, very insubstantial, some concern has been expressed over their possible adverse effects on returning US Space Shuttle craft, following missions in high-inclination orbits. Careful consideration to these will, in future, be given by NASA when planning such missions during the summer months in the northern hemisphere. At high re-entry velocities, even the small quantities of ice crystals within noctilucent clouds could prove abrasive to the outer shell of the Space Shuttle (Ridpath, 1989).

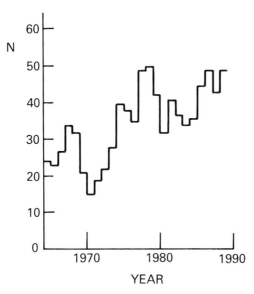

Fig. 9.2. The frequency with which noctilucent cloud displays have been visible from northwest Europe has shown an apparent gradual increase since the 1960s, on which is superimposed the solar-cycle effect. Reproduced from Gadsden, M., *J. Atmos. Terrest. Physics* **52** (4) 247–251 (1990), by kind permission of the author.

9.1.2 Correlations between noctilucent cloud and auroral occurrences

Forming in a layer of the atmosphere not far below the region where the aurora occurs, noctilucent clouds might be expected to show some behavioural correlation with solar–terrestrial activity. It has long been suggested that auroral activity in the overlying thermosphere should sufficiently warm the mesopause that any noctilucent clouds present at the time would be evaporated. Thus, it is expected that the maximum frequency of noctilucent cloud occurrence should be observed at sunspot minimum, when aurorae are uncommon, while few noctilucent cloud displays should be seen around sunspot maximum.

Observations suggest a more complicated inter-relation between the two phenomena. Certainly, large numbers of noctilucent cloud displays were observed from northwest Europe in the solar minimum summer of 1986, including a very bright and extensive display on July 23–24. During the following year, auroral activity stayed low, and another summer with numerous noctilucent cloud displays followed in 1987. The summers of 1988 and 1989, by which time auroral activity was significant, also presented large numbers of noctilucent cloud displays, however, casting some doubt on the reality of the inverse correlation between noctilucent clouds and aurorae. The frequency of noctilucent cloud sightings from northwest Europe did, however, show a substantial drop in the early 1990s. There are some suggestions of a two-year phase lag between auroral activity and the inverse-correlated frequency of noctilucent clouds. The summer of 1995, marked by very low sunspot activity, was again graced with a great many noctilucent cloud displays.

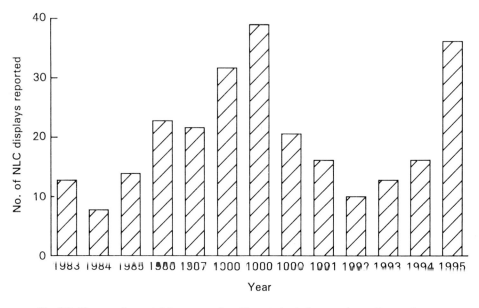

Fig. 9.3. The annual reported frequency of noctilucent clouds from northwest Europe shows a variation roughly two years out of phase with the sunspot cycle. Noctilucent clouds are most common at times of low solar activity.

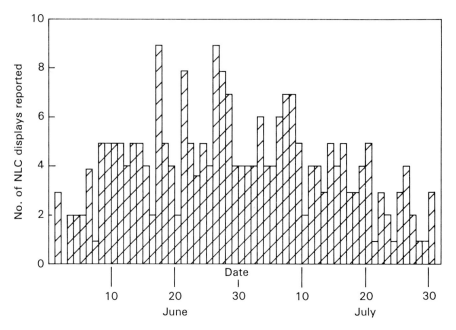

Fig. 9.4. The dates on which noctilucent clouds are most often reported from northwest Europe tend to be in the last 10 days of June and the first week or so of July.

Yet further complications are introduced by simultaneous sightings of noctilucent clouds with a backdrop of auroral activity! Clearly, observations of both phenomena during the summer months will be of increasing value in attempts to unravel the complex interactions taking place in the outer fringes of the Earth's atmosphere.

It is quite likely that the frequency of noctilucent cloud formation is influenced more by the overall level of solar activity at X-ray and ultraviolet wavelengths, than by auroral activity *per se*. As already discussed, the X-ray and UV flux from the Sun plays a significant role in heating of the upper atmosphere (section 8.1). Active regions emitting this short-wave radiation need not be in the correct alignment for the production of auroral activity (section 5.4.2.2) in order to have an effect; flares occurring close to the solar limb are as efficient in heating the upper atmosphere as those close to the Sun's central meridian, though only the latter may be followed by auroral activity.

Fig. 9.3 presents a summary of total noctilucent cloud sightings from the United Kingdom between 1981 and 1995. The rising frequency towards the end of the period coincides with falling sunspot numbers and the approach of solar minimum. Fig. 9.4 shows sightings of noctilucent clouds over these years as a function of date. In general, the best time to see noctilucent clouds from the British Isles appears to be in the last 10 days of June, and into the first week of July.

9.1.3 Appearance and behaviour

Noctilucent clouds bear a superficial resemblance to cirrus, but closer examination reveals diagnostic differences in appearance between the two cloudforms. To begin with, by the time noctilucent clouds are becoming visible on a summer evening, cirrus clouds lying in the same part of the observer's sky should be entering the Earth's shadow and becoming dark. Noctilucent clouds often show a distinctive delicate silvery-blue colour, shading off to gold as a result of atmospheric reddening near the horizon. Highly banded structures are common in noctilucent clouds, and considerable fine detail, including a characteristic 'herring-bone' pattern, is often revealed by binocular examination. The westwards drift of noctilucent cloud features contrasts with the typical eastwards movement of tropospheric weather systems at the latitudes where noctilucent clouds may be observed.

Noctilucent clouds first become visible once the Sun has set below 6 degrees under the observer's westward horizon, and at this time may cover much of the sky. As the depression of the Sun below the horizon increases towards midnight, the noctilucent cloud field will fade somewhat, and diminish in apparent extent. Where the Sun reaches more than 16 degrees below the horizon, noctilucent clouds will fade out altogether. The brightest area in a display normally lies directly above the location of the set Sun, and moves along the poleward horizon as the night goes on. With the approach of dawn, the display can again become more extensive as the Sun rises. When the Sun is less than 6 degrees below the eastern horizon, the sky becomes too bright, and noctilucent cloud forms will appear to melt into the background.

Fig. 9.5 provides visibility curves for noctilucent clouds during their summer 'season' at latitudes of 50°N and 56°N, corresponding roughly to southern England or the northern United States and Canada, and the more favourably placed central Scotland and southern Scandinavia respectively.

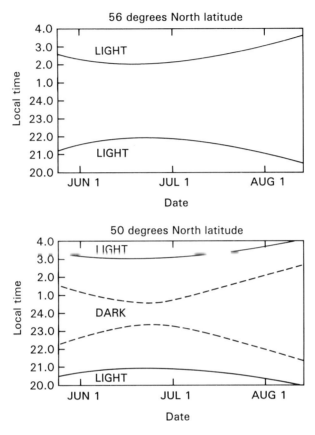

Fig. 9.5. Visibility curves for noctilucent clouds for 50°N and 56°N. Noctilucent clouds cannot be seen when the sky is either too light or too dark, as indicated.

While occasional large displays of noctilucent clouds can fill the whole sky, the brightest regions of the cloud field normally appear in the sunwards half of the sky. The particles which comprise noctilucent clouds are efficient in scattering sunlight forwards, so that any cloud forms overhead and in the half of the sky opposite the Sun will appear rather faint.

Parallactic photography of noctilucent clouds has been carried out by a number of groups, including Scottish amateur astronomers under the guidance of Dr Michael Gadsden of Aberdeen University (Gadsden and Taylor, 1994a, 1994b). Examination of the simultaneous photographs obtained during such work allows precise measurements to be made of noctilucent cloud movements. Noctilucent clouds are carried westwards by a high atmospheric circulation at velocities of up to 400 km h^{-1}. This circulation may have its origin in collisions between particles carried along in the auroral electrojets (section 5.3) and particles in the low ionosphere.

9.1.4 Visual observation
Simple observations of noctilucent clouds remain of interest and value to scientists studying processes in the high atmosphere. In particular, amateur astronomers appear to have

collected the only significant numbers of systematic noctilucent cloud observations throughout the 1970s and early 1980s. These observations may prove their value in assessing whether long-term changes in the frequency of noctilucent clouds really have occurred over this period. This work, in turn, may have important implications with respect to possible man-made changes in the lower atmosphere.

The Balfour Stewart Laboratory of Edinburgh University was, for many years, the main receiving centre for European noctilucent cloud observations. On the Laboratory's closure, responsibility for collecting such observations passed to the Aurora Section of the British Astronomical Association, which in turn provides the observations for archiving at Aberdeen University. Annual reports of all northwest European observations up to 1992 were presented in *Meteorological Magazine* (for example, Gavine 1989), and are now published in the *Journal of the British Astronomical Association*. Among the active groups contributing data are observers in Denmark, who are ideally placed to record noctilucent clouds. Finnish observers are similarly well positioned to see both aurorae and noctilucent clouds, and summarize their results in the magazine *Ursa Minor* published by the Ursa Astronomical Association. Starting in the late 1980s, systematic records of noctilucent clouds visible from Germany have been collected by the Arbeitskreis Meteore (AKM).

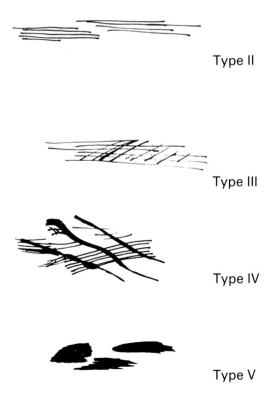

Type II

Type III

Type IV

Type V

Fig. 9.6. Like the aurora, noctilucent clouds show a range of typical forms. Type I, not shown here, is a featureless background veil.

In Canada and the northern United States, observers have become increasingly aware of the importance of recording noctilucent cloud sightings, through the NLC CAN-AM network organized by Mark Zalcik (Zalcik, 1994). Annual summaries of North American observations are published in the *Climatological Bulletin.*

To be of use, observations should be made by standardized methods. As with the aurora, there is a range of typical forms shown by noctilucent clouds, which can be concisely described in observer reports. Fig. 9.6 shows the typical noctilucent cloud structures; standard descriptions used by the British Astronomical Association are listed in Table 9.1.

Table 9.1. Noctilucent Cloud Forms

Type I	**Veil** : very tenuous, lacking in structure, often a 'background' to other forms.
Type II	**Bands**: long streaks often in groups, parallel or crossing at small angles.
Type III	**Billows**: closely spaced, resemble waves or ripples, herring-bone structure very characteristic of noctilucent clouds.
Type IV	**Whirls**: large-scale looped structure, often as complete or partial rings.
Type V	**Amorphous**: similar to veils in the lack of structure, but brighter, usually in patches.

The most fundamental report which can be made is to note that noctilucent cloud was visible from a given location on a particular night. As with auroral observations, the latitude and longitude of the observing station should be given. Double dates should be used to avoid ambiguity—for example, June 27–28, meaning the night of June 27 to the morning of June 28. Universal Time (UT = GMT) should always be used for both noctilucent cloud and aurora observations.

Noctilucent cloud displays tend to change less rapidly than aurorae, so that records of the display's condition at 15-minute intervals are normally sufficient. In order to maximize the numbers of directly comparable reports of a given display, records should ideally be taken on the hour, half hour, and intermediate quarter hours. Observers should note which types of noctilucent cloud structures are present at a given time, and make estimates of the forms' extent in altitude and azimuth. A simple alidade may improve the accuracy with which such measurements can be made. Annotated sketches are often useful in clearly indicating the behaviour of a display. Fig. 9.7 gives an example from the author's observing logbook.

Estimates of the brightness of individual features may be helpful to identification when attempts are subsequently made to combine noctilucent cloud observations from several locations. The brightness of noctilucent clouds is measured on a three-point scale from 1 (very faint) to 3 (bright).

Observers should, naturally, take great care to avoid mis-identifying cirrus or other tropospheric forms for noctilucent clouds. As a general rule, from the latitudes of northwest Europe and Canada, noctilucent clouds will frequently appear in the region of sky below the bright star Capella: by the time that Capella is clearly visible, cirrus in that region of the sky should certainly be in the Earth's shadow. The distinctive colour and structure of noctilucent clouds also aid identification.

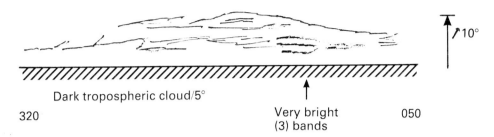

Dark tropospheric cloud/5°

320   Very bright   050
                                                 (3) bands

Fig. 9.7. A useful means of recording noctilucent cloud displays is to make annotated sketches of their appearance at standard 15-minute intervals. This example is taken from the Author's observing log, and depicts a display seen from Edinburgh at 2345 UT on July 15–16 1983.

Noctilucent clouds should not be confused with *nacreous clouds* (also known as polar stratospheric clouds), which occur lower in the atmosphere at altitudes between 21 and 30 km. These are, in any case, most frequently seen at high latitudes during the winter months (McIntosh, 1972; Toon and Turco, 1991). Nacreous clouds, so named for the delicate mother-of-pearl colours which they sometimes exhibit, can be frequently observed from Arctic and Antarctic regions, and have been implicated in processes of ozone depletion.

Observational reports of noctilucent clouds are welcomed by those organizations which collect auroral sightings. Reports of nights when the observer can say with certainty that there was *no* noctilucent cloud visible from a given location are also of value in assessing long-term patterns of noctilucent cloud behaviour.

9.1.5 Photographic observation

As with the aurora, photography provides the means for rapid, accurate recording of the appearance of noctilucent cloud displays. Results on colour slide films can be very attractive. Since the features to be photographed often lie within 15–20 degrees of the horizon, and the illumination of twilight is quite considerable on many nights when noctilucent clouds are present, the photographer is almost obliged to include features of foreground interest, which can also help to illustrate the scale of the display on subsequent viewing. Observers have enjoyed good results with Kodak and Fuji slide films. As a general rule, ISO 400 film works well, with exposures between 2 and 4 seconds, depending on the background sky brightness, at a lens aperture of $f/2.8$.

Since noctilucent clouds will often, from lower latitudes at least, appear largely within a ten-degree strip above the horizon, the best results will usually be obtained with standard 50-mm focal-length camera lenses, rather than shorter-focus wide angle types. The reduced image scale of the latter means that, although the wide expanse of a display may be recorded in its entirety, little detail is seen on the eventual photograph. Rather more effectively, Danish observers have prepared 'mosaics' of several overlapping 50-mm exposures. Only when displays are very extensive is a wide angle lens likely to be more useful for recording noctilucent clouds.

Parallactic photography programmes, using pairs of photographs taken by observers separated by tens of kilometres, have great potential for further elucidating high atmosphere circulation patterns. Dr Gadsden suggests that observers at all locations should take photographs of noctilucent cloud features exactly on the hour, quarter hours and half hour, on nights when noctilucent cloud is present. The use of fixed mounting brackets at each station allows the aiming direction of the camera to be precisely determined. Such a system can also be used for auroral photography.

Simultaneous occurrences of noctilucent clouds and aurora are infrequent, but not unknown. The detailed effects of auroral activity on noctilucent clouds have yet to be closely studied, thanks mainly to a shortage of observational material. Series of photographs taken on those occasions when noctilucent clouds and aurorae appear together may therefore be of great scientific value. Such observations may show, for example, whether the noctilucent cloud field does indeed become disrupted by possible auroral heating under disturbed geomagnetic conditions.

Also of interest are instances of noctilucent clouds appearing at unusually low latitudes (below 50°N, for example). Several such displays were photographed by Jay Brausch from latitude 48°N in Dakota during the summer of 1995.

9.2 ROCKET LAUNCHES AND RELEASES FROM SATELLITES

Occasional anomalous events are reported, where observers at one locality believe they have seen an auroral display, which cannot be confirmed by reports from elsewhere or by measured geomagnetic activity. Some of these may be attributed to local light pollution, as in the case of what appeared to be a display of auroral rays from south London in August 1988, produced by upwards-pointing searchlights illuminating haze. It may sometimes be less easy to account for other events.

The appearance of a 'metallic silver arc-like glow', rising higher into the sky from the north and followed by a second possible arc, as seen from Tank in the Northwest Frontier Province of Pakistan on 1984 December 26 provides an example of the less easily explicable events which are reported from time to time (Livesey, 1985). Initially, it was supposed that this was the manifestation of disturbed auroral conditions, perhaps associated with a persistent coronal hole stream which had generated reasonably active aurorae to mid-latitudes in previous 27-day cycles. Observers at higher geomagnetic latitudes, including northern Scotland and Scandinavia, reported only quiet aurora (arcs and weak rays) on this date, however, while records show that geomagnetic conditions remained relatively undisturbed until some days later. It is thought that a number of such sightings might actually result from gas releases from orbiting satellites or experimental rockets.

Unusual noctilucent cloud effects may also be associated with rocket launches. A brilliant and multicoloured display observed from Finland and Estonia in July 1988 may have been seeded by the exhaust from a Soviet rocket launch (Livesey, 1989). Effects produced by rocket launches are frequently reported by Finnish amateur astronomers.

Observers in Germany and eastern Europe witnessed a bright diffuse glow produced by solar excitation of exhaust from a US Centaur rocket stage on the night of May 3 1994. The V-shaped object, which appeared as bright as magnitude −4 (equal in brilliance to the planet Venus) to some observers, initially sparked some 'UFO' alerts, before its cause was realized (Bone, 1994; Hurst, 1994).

Studies of the upper atmosphere, ionosphere and inner magnetosphere have, since 1955, been carried out using vapour releases from sounding rockets and satellites (Haerendel, 1987). Barium is a favoured material for this work thanks to its low ionization potential. After release, the barium vapour rapidly becomes photoionized by solar ultraviolet. This gives rise to two principal stages in the development of the cloud as observed from the ground. Initially, neutral barium appears as an expanding spherical cloud, emitting by resonance scattering (section 6.2) at 553.3 nm in the green. Photoionization leads to violet emissions at 493.4 and 455.4 nm wavelengths. The cloud then becomes elongated as the barium ions spread along (but not *across*) geomagnetic field lines. This striated form can appear similar to auroral rays.

The development of barium clouds following release can be monitored using low-light television systems and photography. The violet emissions can be recorded on colour film, but may more typically appear dull grey to the eye.

Barium clouds may be quite readily seen with the naked eye, provided the Sun is at least 8 degrees below the observer's horizon while remaining above the horizon at the atmospheric level where the release has occurred. Most barium release experiments have been carried out in the ionosphere between 140 and 400 km altitude. Barium cloud tracing of the ionosphere overlying existing auroral arcs during displays allows investigation of the region in which auroral electrons undergo their final acceleration. While barium cloud structures can resemble auroral arcs, the particles within the former are much less energetic.

Barium releases have occasionally been made over more populous areas where aurorae might also be visible, but advance warning has usually been given in the scientific literature, enabling amateur observers to add their support in monitoring cloud development. A number of barium cloud releases have been made from rockets launched at the Poker Flat range in Alaska. Releases from sounding rockets launched from the Hebridean island of South Uist in the early summer of 1973, for example, were followed by amateur observers in central Scotland. A group of observers from the Edinburgh University Astronomical Society saw the development of a release on the night of May 15 1973 (Gaskell, 1973):

> The barium cloud was first observed at roughly 22.10, some 7° p Castor—a pale yellow disc of about mag −8. It expanded rapidly—taking a mottled appearance after a few minutes. It then merged into an ellipse with major axis running N–S, its size being about 5° × 2° and lengthening with time. The [barium] cloud was observed through passing cloud, and had an overall E–W drift.

Observing from Fort Augustus, Dave Gavine recorded a further barium release on May 16 1973:

> 2342 UT Bright green oval 3° × 3°, approx. 36° elevation in north-northwest

In the late 1960s and early 1970s, experiments were carried out using rocket-borne electron accelerators to trace out geomagnetic field lines. Pulses of electrons, accelerated to energies of 10 keV (similar to those found in aurorae) produced sub-visual excitation of the atmosphere, detectable using low-light television cameras. Electron beams have been successfully fired from one conjugate point to the other with minimal energy loss. Shaped-charge barium releases have also been used to trace geomagnetic field lines, again with the aid of low-light cameras.

Experimental barium releases have also taken place outside the magnetosphere, to investigate plasma motions in the solar wind. The joint US–German–UK Active Magnetospheric Particle Tracer Explorers (AMPTE), consisting of three satellites, generated 'artificial comets' in December 1984 and July 1985. Barium vapour was released from the German Ion Release Module (IRM), monitored from the United Kingdom Subsatellite (UKS). Further work involving gas releases in near-Earth space was carried out by NASA's Combined Release and Radiation Effects Satellite (CRRES) launched in July 1990 (Coates, 1990). Some of the gas releases were observed and photographed by amateur astronomers in the United States during January 1991.

In the 'cold-war' climate of the late 1950s and early 1960s, several nuclear weapon tests were carried out at altitude in the atmosphere by both the Soviet Union and United States. The effects of these test explosions were spectacular, and long-lived. Temporary belts of enhanced particle density and radiation were produced in the plasmasphere region, with deleterious effects for many orbiting satellites, which suffered damage to their solar cells (O'Brien, 1963). Contamination of the magnetosphere by the particles released during these explosions persisted for many years, making measurements of the natural state of the near-Earth environment difficult.

Associated with the explosions of such as Project Starfish, launched from Johnston Island in the Pacific Ocean on July 9, 1962, were auroral-type displays, visible at either conjugate point. The Starfish explosion provided a brief, spectacular display for observers in New Zealand (Boquist and Snyder, 1967).

9.3 AIRGLOW

The night sky of the Earth is not perfectly dark. Even in the absence of auroral activity, a weak background of *airglow* suffuses the sky. Airglow results from re-emission of energy from atmospheric particles following daytime excitation by solar radiation. Light emissions of distinctive character occur during daytime and twilight, too, but the easiest form of airglow for observation is that which occurs at night.

The spectrum of the night-time airglow prominently shows the green (557.7 nm wavelength) line of excited atomic oxygen, but produced by collisions with electrons having energies typically of the order of less than 5 eV (Young, 1966), much lower than that of those giving rise to auroral emissions. Triplets of oxygen atoms combine to produce a molecule of oxygen (O_2) and a single excited oxygen atom, which is responsible for the green line emission. Rocket-borne photometers indicate that this emission occurs in a fairly discrete layer 10 km deep around altitudes of 100 km—that is, at virtually the same height as the base of the auroral layer. Red oxygen emissions at wavelengths of 630.0 and 636.4 nm occur by more complicated mechanisms at greater heights. As with auroral red emissions, these may only occur where particle densities are sufficiently low that quenching as a result of collisions is not a significant process over the relatively long timescales required for the emission to take place.

Emission lines attributable to sodium become prominent in the twilight airglow, as do the red lines of oxygen. The daytime airglow, although 1000 times more intense than the nightglow (Ingham, 1972), is extremely difficult to observe from the ground thanks to the bright sky background. Rocket measurements and the use of extremely narrow-passband filters have allowed dayglow emissions to be isolated from the background for study. Emissions equivalent to those in the nightglow are found at greater intensity in the dayglow, accompanied by more exotic emissions of atomic oxygen and molecular nitrogen which result from collisions between these and free electrons released by solar ultraviolet from other atomic and molecular species.

While weak and diffuse when observed from the ground, airglow can be seen as a fairly strong layer when viewed edge-on in contrast against the blackness of space in photographs from satellites.

Several attempts have been made to simulate the airglow in laboratory conditions. These offer confirmation of proposed airglow mechanisms. Addition of nitric oxide was found to stimulate oxygen emissions, a process subsequently confirmed by rocket releases to generate enhanced local airglows in the high atmosphere (Young, 1980).

The night-time airglow shows variations in brightness, some of which may be associated with geomagnetic activity. Observed brightenings in red oxygen night-time airglow emissions have been attributed to the arrival, via magnetic field lines, of excited particles from the sunlit conjugate point, for example.

REFERENCES

Bone, N. (1994) Is it a comet? Is it a UFO? No it's a rocket stage. *Astronomy Now* **8** (8) 6.

Boquist, W. P., and Snyder, J. W. (1967) Conjugate auroral measurements from the 1962 U.S. high altitude nuclear test series. In: McCormac, B. M. (Ed.), *Aurora and airglow*. Reinhold.

Coates, A. (1990) Surviving radiation in space. *New Scientist* **127** (1726) 42–45.

Gadsden, M. (1989) Noctilucent clouds. *J. Brit. Astron. Assoc.* **99** 210–214.

Gadsden, M., and Schroder, W. (1989) *Noctilucent clouds*. Springer Verlag.

Gadsden, M., and Taylor, M. J. (1994a) Anweisungen fur die photographischen aufnahmen der leuchtenden nachtwolken—103 years on. *J. Atm. Terrest. Physics* **56** (4) 447–459.

Gadsden, M., and Taylor, M. J. (1994b) Measurements of noctilucent cloud heights: a bench mark for changes in the mesosphere. *J. Atm. Terrest. Physics* **56** (4) 461–466.

Gaskell, M. (1973) Barium cloud. *The Astronomer* **10** (110) 21–22.

Gavine, D. M. (1989) Noctilucent clouds over western Europe during 1988. *Meteorological Magazine* **118** 214–216.

Haerendel, G. (1987) Active Plasma Experiments In: Akasofu, S.-I., and Kamide, Y. (Eds), *The Solar Wind and the Earth*. D. Reidel.

Hurst, G. (1994) Fast-moving nebulous object of 1994 May 3. *The Astronomer* **31** (361) 17.

Ingham, W. F. (1972) The spectrum of the airglow. *Scient. Am.* **226** (1) 78–85.

Livesey, R. J. (1985) *J. Brit. Astron. Assoc.* **95** 172.

Livesey, R. J. (1989) BAA Aurora Section *Newsletter* (14) 6.

McIntosh, D. H. (1972) *Meteorological glossary*. H. M. S. O.

O'Brien, B. J. (1963) Radiation belts. *Scient. Am.* **208** (5) 84–96.

Ridpath, I. (1989) *Popular Astronomy* **36** (1) 15.

Soberman, R. K. (1963) Noctilucent clouds. *Scient. Am.* **208** (6) 50–59.

Thomas, G. E., Olivero, J. J., Jensen, E. J., Schroeder, W., and Toon, O. B. (1989) Relation between increasing methane and the presence of ice clouds at the mesopause. *Nature* **338** 490–492.

Toon, O. B., and Turco, R. P. (1991) Polar stratospheric clouds and ozone depletion. *Scient. Am.* **264** (60) 40–47.

Young, L. B. (1980) *Earth's aura.* Penguin.

Young, R. A. (1966) The airglow. *Scient. Am.* **214** (3) 102–110.

Zalcik, M. (1994) In search of noctilucent clouds. *Sky and Telescope* **88** (1) 76–78.

Appendix: Observational organizations

Aurora and noctilucent cloud observations are collected by the British Astronomical Association via its Aurora Section, which also issues standard guidelines for visual observations. The Section can be contacted through:

The British Astronomical Association,
Burlington House,
Piccadilly,
London,
W1V 9AG.

The BAA collects observations from UK observers, and several in the United States: curiously, none of the major amateur astronomical bodies in the US coordinates such work. Many observers in the US do, however, participate in an alert network:

Aurora Alert Hotline,
David Huestis,
25 Manley Drive,
Pascoag, RI 02859, USA.
Tel. (401) 568-9370.

The Finnish Ursa Astronomical Association has an active aurora observing group:

Laivanvarustajakatu 9 C 54,
01400 Helsinki,
Finland.

In Germany, the AKM group has been active in collecting aurora and noctilucent cloud sightings:

Jurgen Rendtel,
AKM,
PF 600118,
D-14401 Potsdam,
Germany.

In the southern hemisphere, observations can be sent to the Royal Astronomical Society of New Zealand:

Royal Astronomical Society of New Zealand,
PO Box 3181,
Wellington,
New Zealand.

North American and Canadian noctilucent cloud observations are collected by:

Mark Zalcik,
9022-132 A Ave,
Edmonton, Alberta,
T5E 1B3,
Canada.

Numerous sources of up-to-date observational material exist, of which the best are probably the Newsletters of the respective observational organizations. Reports of activity from within the previous six weeks or so can be found in *The Astronomer*, published monthly. *Sky and Telescope*, *Astronomy Now*, and *Astronomy* all carry reports from time to time, though these can sometimes be lacking in useful observational details such as dates and places!

Radio observers of auroral and other effects may find contact with the national organizations in that field profitable:

Radio Society of Great Britain,
Lambda House,
Cranborne Road,
Potter's Bar,
Herts.,
EN6 3JE.

The RSGB broadcasts forecasts of likely radio auroral conditions for the week ahead on GB2RS each Sunday.

American Radio Relay League,
Newington,
CT 06111,
USA.

Forecasts of geomagnetic activity are broadcast in the United States by WWV on 2.5, 5.0, 10.0, 15.0 and 20.0 MHz, at 18 minutes past each hour. A recording of these forecasts is accessible by telephone: (303) 497-3235.

Glossary

aa index: Measure of the overall level of geomagnetic activity from equivalent (antipodal) locations in opposite hemispheres, based on 12-hour averages.

Airglow: Low-level diffuse light emission across the whole sky, resulting from excitation of atmospheric particles by solar radiation during daytime.

Ap index: Measure of geomagnetic activity based on *magnetometer* records from 12 observatories worldwide. Similar to the *Kp index*, but the Ap index is linear (rather than semi-logarithmic), and based on 24-hour averages.

***ap* index:** A 3-hourly average of geomagnetic activity on a linear scale similar to the *Ap index*. The *ap* index can be broadly related to *Kp index* values.

Arc: 'Rainbow' or arch-shaped form of *discrete aurora*, usually aligned east–west along the *auroral oval*. From lower latitudes, arcs are seen to have their highest point around the direction towards the magnetic pole. Arcs may be *homogeneous* or *rayed*. The lower edges of arcs arcs are usually more sharply defined than the upper.

Auroral breakup: Main phase of vigorous auroral activity, dominated by moving *arcs*, *bands* and *rayed* forms, during a *substorm*.

Auroral Kilometric Radiation (AKR): Radio emission from the aurora at 100 kHz to several hundred kilohertz frequency, detectable from space.

Auroral ovals: Rings of auroral activity surrounding either geomagnetic pole. Aurora is found more or less permanently around these rings, which are displaced towards Earth's night-side, having their greatest equatorwards extent around the midnight point. Under quiet conditions, the auroral ovals have a diameter of 4000–5000 km. Geomagnetic storms cause the ovals to expand markedly, particularly on the night-side.

Auroral potential structure: Thin sheets of positive and negative charge, aligned to magnetic field lines and lying 10 000–20 000 km above auroral heights in the atmosphere. Electrons undergo acceleration along these thin sheets, gaining sufficient energy to produce auroral excitation on impact with oxygen and nitrogen in the upper atmosphere.

Auroral zone: Circular region around either geomagnetic pole traced out by the maximum equatorwards extent of the *auroral ovals*, and within which aurorae are most likely to be seen. The northern hemisphere auroral zone crosses the North Cape of Norway, Iceland, northern Canada, Alaska and Siberia.

Band: Twisted, ribbon-like form of *discrete aurora*, often resulting from folding of an *arc*, and usually aligned east–west along the *auroral oval*. Bands may be *homogeneous* or *rayed*. Formation of, and movement within, bands often signals the onset of increased auroral activity. The lower edges of bands are usually more sharply defined than the upper.

Bartels diagram: A means of presenting auroral activity data in 27-day strips, each corresponding to a single rotation of the Sun on its axis as seen from Earth. Recurrent aurorae due to persistent solar active features (such as *coronal holes*) become evident by their alignment in adjacent strips.

Bow shock: Wave front produced 'upstream' of a body in the *solar wind*, analogous to that produced by a ship ploughing through water. Solar wind plasma is decelerated and deflected by passage through the bow shock, and then flows around the magnetopause. Bow shocks are seen upwind of the major planets, and ahead of comets.

Chromosphere: Inner region of the Sun's atmosphere, visible at total solar eclipses as a red ring. The chromosphere can be routinely studied at the wavelength of hydrogen-alpha light. The chromosphere lies above the *photosphere*, reaching to about 10 000 km above the solar surface. Overlying the chromosphere is the *corona*. Chromospheric temperatures are typically of the order of 10 000 K.

Cleft: Region in the high-latitude dayside of the *magnetosphere* within which magnetic field lines are closely bundled together. The cleft surrounds the narrower *cusp*.

Conjugate points: Locations of equivalent magnetic latitude and longitude in opposite hemispheres.

Corona: (a) The bright inner atmosphere of the Sun, visible during total solar eclipses. Gas in the corona is at extremely high temperatures (1 000 000 K). (b) Form assumed by the aurora when features pass overhead from the observer's location on the ground. As a result of perspective, rays and other features appear to fan out from a central point.

Coronal hole: Region of open magnetic field in the Sun's inner atmosphere, from which a high-speed particle stream emerges into the *solar wind*. Coronal holes and their associated streams may last for several months, giving rise to recurrent geomagnetic disturbances at 27-day intervals.

Coronal Mass Ejection (CME):   Expanding 'bubble' structure, seen moving outwards through the Sun's *corona* at times of high solar activity, believed to be the result of material being thrown out from the inner solar atmosphere following magnetic reconnection. CMEs are thought to be associated with *solar flares* and the disappearance of *filaments*.

Coronal transient: Equivalent to *coronal mass ejection*.

Crotchet: 'Saw-tooth' feature produced on a *magnetogram* as a consequence of dayside *Sudden Ionospheric Disturbance* events. The brief increase in *D-region* ionization induces an abrupt shift in the ground-level magnetic field, followed by fairly rapid recovery over an hour or so.

Cusp: Narrow high-latitude region on the dayside of the *magnetosphere*, where terrestrial magnetic field lines are closely bundled together and dip sharply towards Earth's surface. *Solar wind* plasma is able to penetrate the magnetosphere via the cusps in either hemisphere.

D region: Lowest part of the *ionosphere*, between altitudes of 65–80 km.

Diffuse aurora: Structureless auroral light which may accompany *discrete* forms.

Discrete aurora: Clearly identifiable auroral structures, such as *arcs* or *bands*, which may be *rayed* or *homogeneous*.

E-cross-B drift: Motion of a charged particle perpendicular to both electrical current and magnetic field.

E-layer: Region of the *ionosphere* around 110 km altitude.

Earth-radii: A convenient unit of measurement for describing distances within Earth's *magnetosphere*. Earth's equatorial radius is 6370 km.

Electrojets: Electron currents flowing eastwards in the evening sector, and westwards in the morning sector, of the *auroral ovals* at altitudes around 100 km. Where these meet, around the midnight point, a region of turbulence (the *Harang discontinuity*) is produced.

F-layer: The highest part of the *ionosphere*, showing a split into two parts. The F_1 layer at 160 km altitude is a permanent feature, while the F_2 layer around 300 km altitude shows a diurnal variation, disappearing at night.

Filament: Equivalent to a *prominence* seen dark by contrast in hydrogen-alpha light against the Sun's disk while in transit. Filaments may disappear following *solar flare* activity elsewhere on the Sun, leading to mass ejections into the *solar wind*. Disappearing filaments may give rise to enhanced geomagnetic activity.

Flaming: Very rapid brightness variation in aurora, wherein waves of brightening sweep upwards from the horizon to the top of the display at a rate of several times per second. Flaming may be the prelude to formation of a *corona* in a very major display, but is also common in the declining phase of a short outburst.

'Flash aurora':   As yet poorly understood phenomenon, reported by several experienced observers, where auroral *arcs*, *bands* or other structures appear in the sky for a matter of only a few seconds before disappearing again.

Forbidden transitions: Electron transitions to energy levels surrounding an atomic nucleus which are not normally permitted by quantum mechanical rules. Excitation during auroral conditions leads to forbidden transitions in oxygen and nitrogen.

Forbush decrease: Interlude of diminished galactic cosmic ray flux at Earth, resulting from the passage of a *coronal mass ejection*.

Geocorona: Cloud of neutral hydrogen from the upper atmosphere, surrounding the Earth to a distance of a few *Earth-radii*.

Geomagnetic latitude: Latitude of a given location with respect to the nearer of the two geomagnetic poles. As a result of the offset of Earth's magnetic and geographical axes, geomagnetic latitude is not equivalent to geographical latitude.

Geomagnetic storm: Period during which a major disturbance of the terrestrial magnetic field is brought about following a *solar flare* or *coronal mass ejection.* Magnetic energy and plasma from the *solar wind* disturbs the magnetotail plasma population, leading to injection of accelerated particles into the upper atmosphere, and an intensification of auroral activity. A major geomagnetic storm may last several days. Such events, which bring the aurora to lower latitudes during expansion of the *auroral ovals,* are most common at times of high sunspot activity.

Ground Level Enhancement (GLE):   Interlude of increased solar cosmic ray flux following a *solar flare* or *coronal mass ejection.*

Ground state: Minimum-energy configuration of electrons surrounding an atomic nucleus.

Harang discontinuity: Turbulent region around the midnight point on the *auroral oval* where the eastwards and westwards *electrojets* meet. Poor short-wave radio communication conditions prevail when a given operating station is closest to the auroral oval at midnight.

Heliopause: Boundary between the *heliosphere* and interstellar space.

Heliosphere: Volume of space within which the solar magnetic field and plasma is dominant, extending perhaps 110–160 AU from the Sun, and thus encompassing the realm of the known planets.

Homogeneous: Description given to either *discrete* or *diffuse aurora* in which no internal structure is evident.

Interplanetary Magnetic Field (IMF):   Magnetic field, whose characteristics are defined by that of features at the Sun's surface, carried by the *solar wind.* The IMF has a typical intensity of 5 nT at the orbit of the Earth. When the IMF polarity is turned southwards with respect to the ecliptic plane, reconnection between the solar wind and Earth's *magnetosphere* is most efficient, leading to enhanced geomagnetic and auroral activity.

Ionosphere: That part of Earth's upper atmosphere between about 60–300 km altitude in which layers of ionization (produced by solar radiation) are found.

Isochasm: Line connecting geographical locations which enjoy the same average annual frequency of aurorae.

Kp index: Measure of the overall level of geomagnetic activity, based on magnetometer readings from 12 observatories worldwide. Averages are given for 3-hour intervals. The Kp index is semi-logarithmic. Kp index <5 corresponds to quiet conditions, Kp > 5 indicates storm conditions.

Lorentz force: Force operating on charged particles in a magnetic field, whereby these are deflected at right-angles to the field, and at right-angles to their previous direction of motion. This leads to spiral trajectories for protons and electrons travelling along magnetic field lines.

Magnetic sectors: Semi-permanent large-scale regions within the *solar wind*, separated by the *neutral sheet*, within which the *Interplanetary Magnetic Field* points either towards or away from the Sun.

Magnetogram: A trace of variations in the local magnetic field as a function of time, as recorded by a *magnetometer*.

Magnetometer: A device for measuring variations in the strength (H in the horizontal direction, Z in the vertical) or the angular deviation (D) of the local magnetic field. Enhanced geomagnetic/auroral activity can be detected by virtue of marked variations in H, Z or D resulting from currents in the *ionosphere*.

Magnetopause: Outer boundary of a planet's *magnetosphere*.

Magnetosheath: Plasma-containing region surrounding Earth's *magnetosphere* between the *magnetopause* and the *bow-shock*.

Magnetosphere: Comet-shaped volume of space in which a planet's magnetic field is dominant over that of the surrounding *solar wind*.

Magnetotail: The extended 'downwind' section of a planet's *magnetosphere*. Earth's magnetotail reaches far beyond the orbit of the Moon on the planet's night-side.

Maunder Minimum: Period of apparently diminished sunspot (and auroral) activity between about 1645 and 1715.

Mirror point: Point in the trajectory of a charged particle moving along a magnetic field at which its motion is exactly perpendicular to the field line. At this point, the charged particle is deflected back along the field line in the opposite direction. Particles in the *Van Allen belts*, for example, are deflected back and forth between mirror points in opposite hemispheres, thereby being 'trapped' in the magnetosphere.

Neutral sheet: A feature of the extended solar atmosphere lying roughly in the Sun's equatorial plane, separating hemispheres of north and south magnetic polarity. Particularly at times of high solar activity, this may become 'pleated', rather than lying in a flat plane.

Noctilucent clouds: Tenuous clouds, possibly comprising water ice condensed onto small particles of meteoric origin, forming close to the mesopause around 82 km altitude. Noctilucent clouds are a summer phenomenon, seen under twilit conditions when the Sun lies between 6° and 16° below the observer's horizon, from latitudes higher than 50°.

Photosphere: The bright visible surface of the Sun. Photospheric temperatures are typically of the order of 6000 K, being reduced to around 4000 K in sunspots, which appear dark in contrast.

Plasma: State of matter in which a gas is completely ionized.

Plasma mantle: Population of plasma flowing along the outside of the lobes in Earth's *magnetotail* on the boundary of the *magnetosheath*.

Plasma sheet: Region of the Earth's *magnetosphere* lying between the opposed north and south magnetic polarity lobes of the *magnetotail*. *Plasma* near the plasma sheet may be injected into the upper atmosphere, causing aurora, at times of high geomagnetic activity.

Plasmasphere: Region of Earth's magnetosphere, containing low-energy plasma lying under the outer *Van Allen belt* to a distance of 4 *Earth-radii*. This plasma population is derived from material lost from the *ionosphere*.

Plasmoids: Pockets of *plasma* from the magnetosphere, enclosed by magnetic 'cage' structures and ejected down the *magnetotail* into the *solar wind* at times of enhanced geomagnetic activity.

Polar Cap Absorption (PCA):   Auroral event occurring polewards of the *auroral oval*, resulting from the arrival of high-energy protons in the *solar wind*. Associated with PCA is disruption of short-wave radio communication at high latitudes, and weak (often sub-visual), diffuse *polar glow aurora*.

Polar glow aurora: Weak, diffuse auroral activity present within the *auroral ovals* during *Polar Cap Absorption* events.

Prominence: Loop of gas from the *chromosphere* reaching perhaps as much as 50 000 km above the Sun's surface, and into the *corona*: prominences are sometimes visible during total solar eclipses as red 'flames' surrounding the dark body of the Moon. Prominences are held up by the solar magnetic field. In the light of hydrogen-alpha, prominences may also be seen in transit as dark *filaments* against the bright background.

Pulsation: Slow brightness variation within aurora, taking place over timescales from a few seconds to several minutes.

Radio aurora: Enhanced short-wave communication conditions, resulting from increased ionization in the *E-layer* during a geomagnetic disturbance. Radio auroral events do not always coincide with visual displays, and vice versa.

Rayed: Description given to forms of *discrete aurora* in which vertical structure is present. Rays may extend for some considerable distance above the base of an *arc* or *band*. Isolated bundles of rays, bearing resemblance to searchlight beams, may also be seen on occasion.

Resonance scattering: Absorption, by a molecule or atom, of sunlight at specific wavelengths and its subsequent re-emission at the same wavelength. Found, for example, in sunlit aurora, where ionized molecular nitrogen (N_2^+) absorbs and re-emits solar radiation at 391.4 nm, resulting in enhancement of this blue-purple emission.

Sector boundary crossing: Interval during which Earth passes through the pleated *neutral sheet* in the *solar wind* from a region of one magnetic polarity to a region of opposed polarity. Sector boundary crossings may add energy to the auroral dynamo, giving a brief enhancement in activity.

Solar constant: Total energy output of the Sun, as measured from satellites above the atmosphere (and found to be anything *but* constant!).

Solar flare: Violent outburst, involving the release of magnetic energy, occurring in the inner solar atmosphere above sunspot groups. Flares are accompanied by ejection of material into the *solar wind*, whose arrival at Earth 24–36 hours later may trigger a *geomagnetic storm*. Flares may be observed at the wavelength of hydrogen-alpha light, and also produce bursts of radio noise.

Solar wind: Continuous outflow of ionized gas (plasma) from the Sun. Under quiet conditions, the solar wind flows past Earth at 400 km s^{-1}. Following *coronal mass ejections* or *solar flares*, local solar wind velocities of 1000–2000 km s^{-1} may be found. *Coronal holes* introduce streams of solar wind flowing at 800 km s^{-1} into the solar wind.

South Atlantic Anomaly: Region above the Atlantic Ocean off the coast of Brazil where the inner *Van Allen belt* reaches a minimum altitude of 250 km above Earth's surface. Artificial satellites passing through this experience increased exposure to energetic particles.

Sporadic E: Thin sheets of enhanced ionization in the *E-layer* of the *ionosphere*, frequently found during daytime in the summer. Sporadic E causes disruption of short-wave radio communication over areas of 1000–2000 km. The mechanisms by which Sporadic E arises are as yet poorly understood.

Stable Auroral Red (SAR) arcs: Weak (typically sub-visual) red aurora produced at mid-latitudes by particles from the *neutral sheet* reaching into the *plasmasphere*, occurring at altitudes of 300–700 km and distinct from the *auroral oval* population.

Substorm: Disturbance of the *auroral ovals* resulting in increased energy input from the *magnetotail*. During substorms, the oval brightens, then expands, on the night-side. Substorms may occur 1–3 times per day under normal conditions, and more frequently at times of high solar activity.

Sudden Ionospheric Disturbance (SID): Rapid increase in the level of ionospheric *D region* ionization resulting from X-ray emissions associated with *solar flares*. The increased ionization causes short-wave radio fadeouts, lasting of the order of a few minutes to an hour, on Earth's dayside.

Sudden Storm Commencement (SSC): Abrupt onset of disturbed geomagnetic conditions, marking the arrival of an interplanetary shock wave associated with a *solar flare* or *coronal mass ejection*. During SSC, the horizontal field strength measured by a *magnetometer* may intensify markedly as a result of compression of the *magnetosphere*. The local geomagnetic field may also show a pronounced angular deviation. SSC events are often, but not always, followed by *geomagnetic storms*.

Theta aurora: Configuration of the *auroral oval* at times of low geomagnetic activity, when a transpolar arc of *discrete aurora* links the day-and night-sides of the oval along the noon–midnight line. So called for its resemblance to the Greek letter.

Van Allen belts: Regions within Earth's *magnetosphere* in which energetic particles are trapped. The inner belt has an equatorial distance of 1.5 *Earth-radii* and contains protons and electrons of both terrestrial and solar wind origin. The outer belt is populated mainly by elctrons from the solar wind, and has an equatorial distance of 4.5 Earth-radii.

Zodiacal light: Diffuse glow, seen as a cone of faint light extending along the ecliptic in the late evening or pre-dawn, resulting from the reflection of sunlight from small particles in the plane of the solar system. The zodiacal light's brightness may be enhanced by energetic particle streams in the *solar wind*.

Bibliography

Where possible, specific references have been provided throughout the text, covering aspects of the aurora. Those wishing to delve deeper may find the references which follow of help. In particular, the works edited by Meng, Rycroft and Frank, and Akasofu and Kamide are rather technical. Eather's enormous book is now quite hard to find. Davis' guide to auroral observing is particularly appropriate for those at Alaskan latitudes.

Akasofu, S.-I, and Kamide, Y. (Eds) (1987) *The solar wind and the Earth*. D. Reidel.

Brekke, A., and Eggeland, A. (1994) *The northern lights, their heritage and science*. Grohndahl and Dreyers Forlag.

Davis, N. (1992) *The aurora watcher's handbook*. University of Alaska Press.

Eather, R. A. (1980) *Majestic lights: the aurora in science, history and the arts*. American Geophysical Union.

Gadsden, M., and Schroder, W. (1989) *Noctilucent clouds*. Springer.

Lang, K. R. (1995) *Sun, Earth and sky*. Springer.

Meng, C. -I., Rycroft, M. J., and Frank, L. A. (Eds) (1991) *Auroral Physics*. Cambridge University Press.

Newton, C. (1991) *Radio auroras*. Radio Society of Great Britain.

Phillips, K. J. H. (1991) *Guide to the Sun*. Cambridge University Press.

Wentzel, D. T. (1989) *The restless Sun*. Smithsonian University Press.

Index